D0904853

The Impact of
Cybernation Technology on
Black Automotive Workers in the U.S.

Research for Business Decisions, No. 84

Richard N. Farmer, Series Editor

Professor of International Business
Indiana University

Other Titles in This Series

The Impact of
Cybernation Technology on
Black Automotive Workers in the U.S.

by
Samuel D.K. James

UMI RESEARCH PRESS
Ann Arbor, Michigan

Produced and distributed by
UMI Research Press
an imprint of
University Microfilms International
A Xerox Information Resources Company
Ann Arbor, Michigan 48106

Library of Congress Cataloging in Publication Data

James, Samuel D. K., 1943-
  The impact of cybernation technology on Black
automotive workers in the U.S.

  (Research for business decisions ; no. 84)
  Originally presented as the author's thesis (Ph.D.)—
University of Delaware, 1984.
  Bibliography: p.
  Includes index.
  1. Afro-American automobile industry workers—
Effect of technological innovations on.   2. Robotics—
United States.   I. Title.   II. Series.
HD6331.18.A8J36 1985   331.6'3'96073   85-16500
ISBN 0-8357-1719-4 (alk. paper)

In loving memory of my parents, Samuel Eli James and Rosie Phillips James

Contents

List of Figures

List of Tables

Acknowledgments

Warm and sincere gratitude and appreciation are extended to the many individuals who have inspired, encouraged, and assisted me during the course of this study. Robert Warren proved to be a worthy senior scholar as well as a colleague and friend. The late Mark Haskell provided invaluable insights and contributions; and Paul Solano made critical and essential comments. John Byrne, one of the brightest young scholars on the horizon, contributed tremendously to this study. A very special word of appreciation goes to Jeanne Wortham, who typed and retyped the drafts, and to Jennifer Ransom, for her unselfish assistance throughout the study.

1

The Implications of Cybernation Technology for American Society

Public Policy Questions

This research was motivated by three interrelated questions concerning national employment policy. In broad scope they related to the current and future impact of cybernation technology on the United States work force. (See the appendix for definitions of terms used in this study.) More specifically, they focus on its short- and long-run effects upon Black production workers in the automotive industry.

The initial question to be answered is whether or not a cybernation revolution is occurring today because of a merging of automation and cybernetics which could potentially render human labor power and dexterity obsolete in the industrial sector. This technological shift in the mode of production is variously referred to as the cybernetic, the cybernation, or the computer revolution. Its tools are electronic, digital, and analog computers based on the principle of automatic feedback. These tools significantly augment man's mental and physical powers. In doing so, they embody the potential for an age of effortless abundance in which machines of material production operate fully and correct themselves without the direct intervention of a human being. If cybernation is occurring rapidly and extensively enough to make such a future possible, an obvious policy question is whether the resulting benefits would be equitably distributed within society. If this were not the case, a plethora of economic, political, and social problems could arise which would require a complex public response. In the broadest sense, questions of whether cybernation is occurring on a significant scale, and whether potential social and economic problems could be ameliorated, are issues which must be addressed.[1]

Second, assuming that these systematic aspects of cybernation will evolve, one way to explore their implications for the work force is to examine a case where the process is underway. One of the best examples is that of the current robotization of assembly plants in the auto industry. Cybernation technology in the form of the industrial robot has immediate applicability to production-line activities.

As a result, the traditional jobs of unskilled, semiskilled, and even some grades of skilled production workers are at risk of becoming obsolete. By looking at current evidence from the auto industry, it is possible to determine whether assembly-line related workers—and particularly Black workers—are vulnerable to at least short-run obsolescence. Again, assuming that one of the effects of cybernation will be job losses on the production line (and that these effects will fall disproportionately on Blacks), a further question is whether the obsolescence will be of a longer term, if not permanent, nature.

Third, even if there are immediate negative effects on production workers, policy makers tend to assume that displaced workers will eventually be reabsorbed in other sectors of the economy. These assumptions are made without the development of public policies to foster this outcome. Given the nature of cybernation, this conclusion is open to question. For example, as cybernation technology is applied to the industrial sector, the mode of production becomes increasingly capital-intensive instead of labor-intensive. Production methods based on this technology generate their own internal logic. Human input becomes redundant; and in the absence of specific public policy, the results might be massive long-run unemployment if the effects of private corporate decisions are not internalized by the public sector. Whether obsolete workers can be retained and reabsorbed into new production jobs would depend upon the capacity of the major actors involved—labor, management, and government—to properly frame, debate, and respond to the issue through the adoption of a national employment policy.

The case of the auto industry provides an excellent point of entry for research on these interrelated issues, particularly on the way these principal actors have responded to the current and potential effects of cybernation on production workers. Such an inquiry also offers an opportunity for considering alternative ways of treating the general question of cybernation and employment in public policy terms.

Social Policy and Cybernation

An examination of these questions forms the basis of the present study and can now be considered in greater detail. A discussion of the policy issues arising from cybernetics is incomplete without recognizing the contributions made by Norbert Weiner. Though Weiner developed the mathematics necessary for building the cybernetic machine, he never ceased reminding society of the potential negative effects of these artificially intelligent machines on human beings. Merely 10 years after witnessing the industrial application and adoption of these automatic control systems, he warned of the certainty that future machines would possess the power to take many jobs from human beings. Weiner also noted that instead of valuing an individual by the job he does, we should value him as a person.[2]

F. H. George's classic, *Automation, Cybernetics and Society,* contains a description of the automatic factory in which:

all the operations are performed by automatically controlled systems. . . . There must be automatic repair whenever breakdown occurs, and automatic maintenance; there will be automatic testing of the product and quality control introduced into the system, automatic packing, wrapping, storage and delivery.[3]

He goes on to note the delicate balance that exists between the positive and negative effects of automation, and comments that the view he has adopted is "that of the social scientist and logician who is interested in humanity first and last, and who believes that humanity stands to gain new worlds by proper use of scientific advance but is equally close to the possibility of hell-on-earth."[4]

John A. Sharp advises us that the cybernetic revolution cannot be fully grasped by limiting one's analysis to the machine. Rather, one should consider a multisocial view. He also added that short-run considerations of the early stage might include: technological unemployment, underemployment, the social lag, labor mobility, retraining, the growth of big business, and interference. In short, Sharp argues that "Advancing cybernetic technology has created significant threats to institutions which are basic to our democratic and capitalist society."[5]

On the problem of this technology and labor, Sharp has the following to say: "The institutions of our present society are not geared to handle these problems, especially in the areas where cybernation will have its greatest impact. Even labor unions have recognized their insufficiency in dealing with the problems of worker security and structural unemployment."[6]

Another authority on cybernetics, Demczynski, reminds us that

The central fact of the cybernetic revolution is that it is introducing machines which augment our human capacity for rational data processing on a scale analogous to that on which steam, electrical and internal combustion engines augmented our physical powers in the first industrial revolution.[7]

Demczynski goes on to explain that the cybernetic machine can radically change the design and method of manufacturing. Hence "The majority of the mental tasks performed at present by human beings can also be modified so that they can be executed by existing cybernetic machines millions of times faster and much cheaper."[8]

These cybernetic machines can also perform managerial and executive functions, and are valuable research tools. But Demczynski remains uncertain about society's ability to meet the challenge of these machines as far as labor was concerned: "Cybernetic machines can ultimately make a poorly educated man redundant as a producer of goods and services, just as mechanically powered machines made redundant men who had little but physical energy to offer."[9]

Obviously, something is happening to the production relationship which heretofore existed between man and machine. Human labor and/or dexterity is becoming increasingly obsolete as a complement to the mode of production.

Whether one uses the term cybernetic or cybernation, the technology is the same, as is the impact on human beings.

In his writings of 1970, D. N. Michaels (a social scientist who created the term cybernation by combining the terms automation and cybernetics) expressed his belief that within 20 years machines would be able to do jobs involving original thinking.[10]

This technology permits greater rationalization of managerial activities. Michaels assumed that cybernation technology would be required to meet the demands of a growing population and maintain or increase the rate of growth of the gross national product. Moreover, he emphasized that this technology possessed the capability of either emancipating or enslaving mankind. That is, in a competitive society, those organizations and industries that have successfully introduced this technology usually will have economic advantages over those using human beings for equivalent work. As a result of cybernated production, Michaels expected massive blue-collar unemployment in the chemical industry, potentially large-scale unemployment of middle management, and insufficient job growth in the service industries.[11]

Herein lies the crux of the policy problem the United States is facing: the complete understanding and evaluation of the nature of this cybernation revolution whose technological dynamics may render obsolete large portions of the work force.

Short- and Long-Term Effects on the Auto Industry

We know that production machines are capable of replacing human workers. We also know that certain subsets of the American population are dependent upon their "labor power" for economic and employment survival. Black American workers are one of those subsets of workers still relying heavily on production-line jobs, particularly in automobile assembly plants. Consequently, the problems associated with the robotization of those plants will fall heavily and disproportionately on the Black production workers, or to quote James Boggs: "Cybernation . . . is eliminating the 'Negro jobs'. "[12]

The impact of cybernation thus assumes added public policy significance. The long-run reabsorption of these obsolete workers into the economy is an obvious solution. But this might be more difficult than previously anticipated. Namely, from a structural perspective, the United States is changing from a predominately goods producing society to a service producing society. Skill requirements of the work force are also changing. White-collar rather than blue-collar workers will replace those eliminated by cybernation (if, in fact, they are replaced). The question of the long-run reabsorption of obsolete production workers not only depends upon other positions being available, but more importantly, on the ability of these displaced workers to compete for white-collar jobs. Though virtually none of the basic questions about the effects of cybernation of the work force have been answered, this is especially true of the question of how cybernation will affect subgroups in the work force.

Cybernation Technology as a National Issue

There has been recurring interest in the relationship of technological change and the employability of man as a complement to machines in the United States. For example, in the 1930s America was experiencing rapid changes in science and industry. Issues and questions were raised concerning the future character of society and industrial employment. These concerns were eventually placed on the national political agenda and public debate ensued. One of the major analyses of the problem was undertaken by the National Resources Committee, which found (in 1937) that the long-run technological trends would have favorable effects on the economy and employment.[13]

The 1950s and particularly the 1960s can be considered another focal point in the public debate around the issue of industrial technology and employment. During this period emphasis was placed on the issues of whether and to what extent automation or automated production techniques would displace human workers. In other words, a complex set of issues was discussed:

> (1) the extent of possible and probable displacement of personnel, (2) the possible shifts and distortions which may arise in the distribution of mass purchasing power, (3) the equitable distribution of the expected gains in productivity, (4) the effect upon our business structure, (5) the effect upon the volume and regularity of private investment.[14]

Differing views on these issues, especially on the first one, were most clearly expressed in the mid-1960s. One of the most important statements made on technological change and employability of humans came from the work of a group of concerned social scientists and computer experts meeting in conference at the Institute for Advanced Studies, Princeton University.

The Ad Hoc Committee on the Triple Revolution of 1964 (the name by which this group of individuals became known) placed special emphasis on the disruptive aspects of cybernation technology together with its negative impact on the Black American. The concerns of the Ad Hoc Committee were written in a memorandum to the President and subsequently presented at a 1966 Senate hearing.[15]

The memorandum argued that the impact of cybernation technology on both society and the Black American would be negative unless American society established some type of ameliorating institutional framework. Because material production could be carried on without the direct intervention of human workers, the committee argued that Black Americans would be disproportionately penalized and exiled from the production system by cybernation technology.[16] As a solution to these problems, the Ad Hoc Committee developed and elaborated on an extensive set of proposals designed to minimize the social costs of the transition to cybernation technology.

The Ad Hoc Committee was not alone in raising the concerns of the application of cybernation technology to the mode of production and of its impact on the work force in general. As early as the mid-1950s, but especially at the instigation of President Kennedy in 1960, the national government exhibited concern about the impact of automation. Following a series of public hearings in the 1960s (in which the Ad Hoc Committee participated) the establishment of a national commission to investigate the issues of automation, technology, and employment was recommended in H. R. 10310. H. R. 11611 actually established the National Commission.[17]

The Ad Hoc Committee and the National Commission both started from the same major premise, namely, that the new technology would have short-run disruptive effects on the work force. However, their basic assumptions about the long-run consequences and the appropriate public responses were dissimilar to the extent that the perspective of the Ad Hoc Committee can be viewed as revolutionary and heavy (in-kind) and that of the National Commission as evolutionary and normal (in-degree). The former perspective foresaw the need for a basic restructuring of the institutions in society and the meaning of work. The evolutionary perspective reasoned along the lines of an equilibrium model which resolves short-run disequilibrium in the job market through incremental changes within the existing institutions.

Since the mid-1960s, there has been no comparable formal confrontation of the issues in national debate. Ironically, the evidence of emerging changes in the nature of work from the impact of technology is far greater now than it was at that time. This is not to say that the impact of cybernation has been ignored. Rather, the problem has not been fully framed in public discussion and the institutions and processes for doing so have, to this point, remained inadequate. A consideration of how labor, management, and government have dealt with cybernation in the auto industry provides a clear picture of the nature of the dilemma.

Public and Private Responses to the Problem

An insight into this policy paralysis can be gained by examining the principal actors' responses to the application of cybernation technology to the auto industry's production core, its assembly plants. The actors of concern are the auto industry's management, the federal government, and the union—the United Automobile, Aerospace, and Agricultural Implement Workers of America (UAW).

For example, the auto industry's management views cybernation technology— the industrial robot, or computer-aided manufacturing and computer-aided design—as the main way to increase productivity, save on labor costs, and improve the overall management structure of the industry. Computer-aided manufacturing (CAM) and computer-aided design (CAD), both outgrowths of the cybernation revolution, promise to render the American auto industry competitive once again. That is: "If wisely and widely applied, CAD/CAM could overcome the productivity stagnation that has led to questions about the ability of American industry to remain competitive on the world scene."[18]

The quotation above is only the proverbial tip of the iceberg regarding management's perspective. *Automotive Industries* reported in 1977 that General Motors, Ford, and Chrysler, in order to rebuild the industry's profitability, plan to spend 29 billion dollars on equipment and facilities over a five-year period.[19]

Management sees no serious labor-related problems in shifting to cybernation. While robotics will displace workers from the production line, the benefits (they argue) will outweigh the costs. The basis of management's optimism is that as more goods can be produced at a lower price, the demand for labor will rise. This will, in turn, reduce unemployment, expand incomes, and increase sales. Since management envisions no labor-related problem, it offers no comprehensive policy on the issue.

The federal government is aware that there is a potential employment problem accompanying the introduction of cybernation technology into auto production. The plight of the auto industry, for instance, was discussed at the national level in 1975, 1979, and 1980 in Senate and Congressional hearings.

During the 1979 and 1980 Chrysler Loan Guarantee Hearings, the recovery program advocated for the auto industry necessarily involved labor cost reductions. But herein lies the dilemma: on the one hand, the government has supported robotization policies potentially detrimental to the worker, while on the other hand it has advocated policies supportive of those recently displaced. In short, though the government enunciates corrective policies for displaced workers (retraining is a good example), these policies are at best stop-gap and incomplete. Hence, the government has neither developed a systematic policy addressing the displacement that results from the application of cybernation technology, nor has it sought to maintain the discussion of labor-displacing technology as a high priority item on the public agenda. In short, while cybernation technology is supported and encouraged, cushions, training, and retraining—though advocated—are not adequately financed.

Auto workers themselves seem to face the worst dilemma. The UAW views robotic technology as necessary and inevitable. More specifically, while it must support the introduction of this productivity-increasing and labor-saving technology, it must simultaneously support the retention of the jobs of the workers that it represents. UAW leadership has historically accepted and encouraged technological change in the belief that the new production technology would affect few workers. However, only very recently has the new production—robotic technology—begun to seriously concern the UAW and motivate the discussion of policies designed to minimize the displacement effects on its constituency. Ironically, this has occurred at a time when its bargaining power is very poor. The reasons why the dilemmas of cybernetic production are not currently being addressed as a policy issue rest in large part with the situation of the principal actors.

The Objectives and Structure of the Study

The objectives of this study are several. First, it is designed to assess the extent and magnitude of the changes that are occurring because of the substitution of cybernated machines for human labor in industrial production. Second, the public policy questions raised in the United States in the 1960s concerning the effects of cybernation on the labor force in general (and with specific emphasis upon its impact on Black workers) will be assessed in light of present conditions. To do so, a case study was undertaken to evaluate the short-run occupational vulnerability and risk of the Black worker from the application of robots to automotive production. Third, the study will consider the question of the likelihood that displaced Black production workers will be reabsorbed—in comparable jobs—in the long run by the auto industry or the general labor market. This inquiry will necessarily be speculative for two reasons. One is that, other things being equal, valid projection techniques are not available. The other reason involves public policy: the timing, extent, and nature of governmental intervention cannot, at this point, be predicted. Thus, a major issue concerning this third objective is whether the national government will attempt to formulate a policy to mitigate potential negative effects of cybernation on production workers. In addition, the government responses in selected industrial countries to the actual or potential displacement of human labor by robots and related technology will be reviewed.

Because the nature and effects of cybernation are still unfolding and embryonic, the study is concerned with the future as well as the present. Also, since public response is as yet undefinable, the analysis must be largely suggestive. If it can provide adequate data to show that Black auto production workers are at high risk in the short- and long-run in the job market, and can effectively argue that an issue which is of national importance is not receiving adequate public attention, its goals as both an empirical and policy study will be achieved.

With the preceding remarks in mind, we now turn to consider the organization of the study. Chapter 2 expands upon the question of whether there is a cybernation revolution occurring today which has the potential to render manual labor power and/or dexterity obsolete in the industrial sector. The positions of the Ad Hoc Committee on the Triple Revolution and the Commission on Automation, Technology and Employment are examined in detail. The chapter concludes by citing available evidence on the current nature of the impact of cybernation technology worldwide—on society and on the work force.

Chapter 3 deals with the circumstance which has caused cybernation to be introduced into the automobile industry, how the technology is being applied, and the type of jobs that are affected. Chapter 4 begins with a discussion of the significance of the auto industry to the Black American—as a main point-of-entry into the industrial sector, as a source of employment with one of the highest industrial wages in the country, and as an area of employment whose geographical location is relatively close to large concentrations of Blacks. The discussion of the Black worker's *vulnerability* and *risk* to robotic technology is built upon these factors. Chapter 4 concludes with a discussion of the prospects for reabsorbing displaced workers in a changing automotive industry as reflected in Bureau of Labor Statistics (BLS) forecasts and other sources.

Chapter 5 discusses some of the policy dilemmas that the principal actors face as the mode of production in automotive industry shifts from labor power to cybernation (robotic) technology. The main purpose of this chapter is to review the nature of policy responses of the principal actors—the auto industry's management, the government, and the union—to this rapid and radical shift to cybernation technology.

Chapter 6 summarizes the findings of the study, discusses policy implications for the United States of the effects of cybernation on employment, and concludes with suggested future issues and research questions.

2

The National Debate of the 1960s: The Impact of Cybernation Technology on Labor and Industry

That a major scientific and technological revolution is occurring is evident from the tremendous advances made in space technology, military hardware, and industrial production. The aspects of this revolution that impinge directly upon the present study are the dynamics and consequences of cybernation technology. Such technology presents a new mode of computer-aided production, with the potential to render obsolete most forms of manual production labor. It appears to have immediate impact on unskilled and semiskilled production jobs—jobs traditionally held by Black Americans.

With the introduction of cybernation technology into the American industrial sector, national debate heightened regarding such technology's impact on the work force. From one perspective it was argued that this new labor-saving technology had the potential of impacting very heavily and disproportionately on the Black American production worker. In short, it was argued that this effect could also fall very heavily on any subset of the population not possessing the training or skills required to accommodate a cybercultural work environment. These were the concerns that gained the attention of a committee of concerned social scientists and computer experts (herein referred to as the Ad Hoc Committee) who, after a conference in 1963 at the Institute for Advanced Studies, Princeton University, under the direction of J. Robert Openheimer, issued a memorandum entitled "The Triple Revolution." The conclusions reached in this memorandum regarding the impact of cybernation technology on both American society in general and Black Americans in particular were pessimistic.

The concern over the labor-saving effects of technology was also the focus of formal governmental attention during the 1960s. Congress established the National Commission on Technology, Automation, and Economic Progress, which was given the responsibility to consider a number of the same kinds of issues addressed by the Ad Hoc Committee. The National Commission reached a quite different set of conclusions about the dynamics of technology and employment, and about the assumptions which should guide the role of government in the changes that were expected.

Close to two decades have passed since this wide public discussion of technology and employment was sparked by the National Commission and the Ad Hoc Committee. The debate—and even any kind of systematic attention to the issues raised—has virtually ceased. Yet the industrial world has come increasingly under the dictates of cybernation technology; and it has become increasingly important to reassess the merits of the debate of the 1960s.

This chapter will first examine and compare the assumptions, arguments, and policy conclusions of the Ad Hoc Committee and the National Commission. Current literature will then be surveyed to determine the extent to which cybernation is being applied to industrial production, and to evaluate its impact on employment.

The Concept of Revolutionary versus Evolutionary Change in the Structure of the U.S. Economy

An important distinction, central to this chapter, should be noted at the beginning of the discussion. It concerns the concepts of revolutionary change and evolutionary change, with the former term being identified with the Ad Hoc Committee, and the latter with the National Commission.

The term "change" as it is used here refers both to qualitative and quantitative processes. For instance, the period of the mechanization of production required both material and institutional complements to reveal its logic. From the qualitative perspective, man had to conceptually internalize the prerequisites of the new production technology (which was mentally a gigantic step above the premechanized production world). From the quantitative perspective, the mechanization of agriculture resulted in a tenfold increase in output over the previous system.

Because of the nature and impact of the change, we must be able to distinguish such differences. The term "revolutionary change" refers to a change which completely and "irreversibly" modifies both the mode of production and its accompanying superstructure (e.g., its political, economic, and legal institutions). This change is referred to as revolutionary, heavy, and in-kind. Conversely, when technological change is viewed as not modifying "irreversibly" (particularly the superstructure), it is considered evolutionary, or normal and in-degree.

Figure 2-1 provides a graphic framework for discussing these differences. The analysis of this figure begins at point (a), at the base of society, where a superior-inferior causal relationship exists. The forces of technology (cybernation) modify (b), the mode of production (i.e., the organization of society's production system). Both the Ad Hoc Committee and the National Commission assume this causal relationship.

The Ad Hoc Committee sees this process as rapid, abrupt, and creating a disequilibrium. The National Commission, however, views the change as less disruptive, more adaptive, and incremental. This part of the discussion constitutes the first major distinction between the revolutionary versus the evolutionary concepts. Turning to point (c), we see that there is a reciprocal relationship between

Figure 2-1.   A Schematic of the Impact of Cybernation Technology on Society

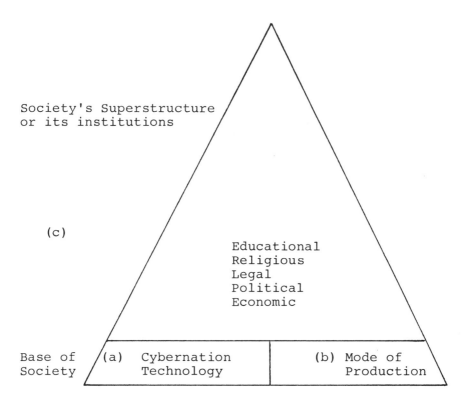

the base of society and its superstructure. From a revolutionary impact perspective, the structure of these institutions must be radically changed before any proposal designed to facilitate the transition (without unacceptable social costs) to a cybernetic mode of production can be successful. Herein lies the second major distinction between the concepts "revolutionary" and "evolutionary." The Ad Hoc Committee called for heavy and in-kind changes in America's industrial system in order to eliminate poverty. The National Commission states that some institutional changes are necessary, but only to the extent that they are normal and in-degree. These two distinctions comprise the crux of the revolutionary and evolutionary public policy positions.

The Ad Hoc Committee's Position on the Impact of Cybernation Technology

In the fall of 1963, a conference was held in Chicago which was primarily attended by bankers and industrialists, including several social scientists who presented papers which were to form the basis of the Ad Hoc Committee's document, "The Triple Revolution."

W. H. Ferry, the organizer of the Ad Hoc Committee, presented a paper outlining his ideas on cybernation technology and the replacement of human beings in production jobs by machines. Robert Theobald also spoke concerning the need for social planning and commitment in view of the implications of cybernetic technology. Ralph Helstein dealt with the subject of automation and worker displacement in the packing-house industry. The common theme pervading these papers was that serious attention was not being given to the possible negative impacts of cybernation technology. Because these ideas were not well received, Mr. Ferry suggested the convening of a conference specifically to discuss the implications of the cybernetic revolution.[1]

In October of 1963 (at the invitation of J. Robert Openheimer, the Director of the Institute for Advanced Studies, Princeton University), the Ad Hoc Committee on the Triple Revolution met. The position reached by the Ad Hoc Committee was that "the traditional link between jobs and incomes is being broken"; and the cause of this break is the cybernation revolution—the combination of the electronic computer with the automated, self-regulating machine. According to the Committee:

> The fundamental problem posed by the cybernation revolution in the United States is that it invalidates the general mechanism so far employed to undergird people's rights as consumers. . . . Up to this time, economic resources have been distributed on the basis of contributions to production. . . . In the developing cybernated system, potentially unlimited output can be achieved by systems of machines which will require little cooperation from human beings.[2]

The Ad Hoc Committee did not argue that cybernation technology itself produced negative effects on labor. Rather, it argued that the social, economic, and political norms of decision-makers—together with the manner in which this technology is introduced as a component of the production process—will determine whether societal costs can be avoided. On March 22, 1964, following the Princeton Conference, the Ad Hoc Committee sent the following memorandum to President Johnson:

> Dear Mr. President:
>
> We enclose a memorandum, "The Triple Revolution," for your consideration. This memorandum was prepared out of a feeling of foreboding about the nation's future. The men and women whose names are signed to it think that neither Americans nor their leaders are aware of the magnitude and acceleration of the changes going on around them. These changes, economic, military, and social, comprise The Triple Revolution. We believe that these changes will compel, in the very near future and whether we like it or not, public measures that move radically beyond any steps now proposed or contemplated.

We commend the spirit prompting the War on Poverty recently announced, and the new commissions on economic dislocation and automation. With deference, this memorandum sets forth the historical and technological reasons why such tactics seem bound to fall short. Radically new circumstances demand radically new strategies.

If policies such as those suggested in "The Triple Revolution" are not adopted we believe that the nation will be thrown into unprecedented economic and social disorder. Our statement is aimed at showing why drastic changes in our economic organization are occurring, their relation to the growing movement for full rights for Negroes, and the minimal public and private measures that appear to us to be required.

Sincerely,

[The following comprised the Ad Hoc Committee:]

Donald A. Agger	Gunnar Myrdal
Dr. Donald B. Armstrong	Gerald Piel
James Boggs	Michael D. Reagan
W. H. Ferry	Ben B. Seligman
Todd Gitlin	Robert Theobald
Roger Hagan	William Worthy
Michael Harrington	David T. Bazelton
Tom Hayden	Maxwell Geismar
Ralph L. Helstein	Philip Green
Dr. Frances W. Herring	Alice Mary Hilton
Brig. Gen. Hugh B. Hester	H. Stuart Hughes
Gerald W. Johnson	Linus Pauling
Irving F. Laucks	John William Ward[3]

Obviously, the committee felt that there was an urgent need for public policies designed to mitigate the negative consequences of this technology. The memorandum set out in detail the reasons for this conclusion. The Ad Hoc Committee began its analysis thusly:

A new era of production has begun. Its principles of organization are as different as those of the industrial era were different from the agricultural. The cybernation revolution has been brought about by the combination of the computer and the automated self-regulating machine. This results in a system of almost unlimited productive capacity which requires progressively less human labor. Cybernation is already reorganizing the economic and social system to meet its own needs.[4]

The committee argued that the negative effects of this "new era" would fall disproportionately on Black Americans:

The Negro claims as a matter of simple justice, his full share in America's economic and social life. He sees adequate employment opportunities as a chief means of attaining this goal. . . . Negroes are the hardest hit of the many groups being exiled from the economy by cybernation.[5]

Further on it was stated that:

The U.S. operates on the thesis, set out in the Employment Act of 1964, that every person will be able to obtain a job if he wishes to do so and that this job will provide him with resources adequate to live and maintain a family decently. Thus job-holding is the general mechanism through which economic resources are distributed. Those without work have access only to a minimal income, hardly sufficient to provide the necessities of life, and enabling those receiving it to function as only "minimum consumers."[6]

The Ad Hoc Committee proceeded to elaborate on its thinking regarding the industrial system in the U.S. and its ability to abolish poverty, pointing out that

There is no question that cybernation does increase the potential for the provision of funds to neglected public sectors. Nor is there any question that cybernation would make possible the abolition of poverty at home and abroad. But the industrial system does not possess any adequate mechanisms to permit these potentials to become realities. The industrial system was designed to produce an ever-increasing quantity of goods as efficiently as possible, and it was assumed that the distribution of the power to purchase these goods would occur almost automatically. The continuance of the income-through-jobs link as the only major mechanism for distributing effective demand—for granting the right to consume—now acts as the main brake on the almost unlimited capacity of a cybernated productive system.[7]

These comments tend to contradict both Say's law of markets (i.e., supply creates its own demand) and Keynes's theoretical model of aggregate demand and fiscal policy, which are silent about the total absence of and the system's inability to generate effective demand (purchasing power). However, Keynes was somewhat aware that the government might be required to introduce measures designed to achieve a much better distribution of resources together with a reduction in unemployment. On this point, the Ad Hoc Committee concluded that

An adequate distribution of the potential abundance of goods and services will be achieved only when it is understood that the major economic problem is not how to increase production but how to distribute the abundance that is the great potential of cybernation. There is an urgent need for a fundamental change in the mechanisms employed to insure consumer rights.[8]

The Ad Hoc Committee then attempted to reconcile a major paradox plaguing the economic system based on cybernation—surplus capacity and unemployment. The underlying cause of excess unemployment, argued the Committee, is the fact that intelligent machines are increasing in capacity faster than human beings. The result of this incongruity is the emergence of a permanent impoverished and jobless social class in the midst of potential abundance.

The postulation that an impoverished and jobless class is being created serves as the most critical point advanced by the Committee; and it is useful to restate in detail the Committee's evidence to support it:

1. The increased efficiency of machine systems is shown in the more rapid increase in productivity per man-hour since 1960, a year that marks the first visible upsurge of the cyber-

nation revolution. In 1961, 1962, and 1963, productivity per man-hour rose at an average pace above 3.5 percent—a rate well above both the historical average and the postwar rate. Companies are finding cybernation more and more attractive. Even at the present early stage of cybernation, costs have already been lowered to a point where the price of a durable machine may be as little as one-third of the current annual wage cost of the worker it replaces. A more rapid rise in the rate of productivity increase per man-hour can be expected from now on.

2. In recent years it has proved to increase demand fast enough to bring about the full use of either men or plant capacities. The task of developing sufficient additional demand promises to become more difficult each year. A $30 billion annual increase in gross national product is now required to prevent unemployment rates from rising. An additional $40 to $60 billion increase would be required to bring unemployment rates down to an acceptable level.

3. The official rate of unemployment has remained at or above 5.5 percent during the sixties. The unemployment rate for teenagers has been rising steadily and now stands around 15 percent. The unemployment rate for Negro teenagers stands about 30 percent. The unemployment rate for teenagers in minority ghettoes sometimes exceeds 50 percent. Unemployment rates for Negroes are regularly more than twice those for whites, whatever their occupation, educational level, age, or sex. The unemployment position for other radical minorities is similarly unfavorable. Unemployment rates in depressed areas often exceed 50 percent.

4. An efficiently functioning industrial system is assumed to provide the great majority of new jobs through the expansion of the private enterprise sector. But well over half of the new jobs created during 1957-62 were in the public sector—predominately in teaching.

5. Cybernation raises the level of the skills of the machine. Secretary of Labor Wirtz has recently stated that the machines being produced today have, on the average, skills equivalent to a high school diploma. If a human being is to compete with such machines, therefore, he must at least possess a high school diploma.

6. A permanently depressed class is developing in the United States. Some 38 million Americans, almost one-fifth of the Nation, still live in poverty. The percentage of total income received by the poorest 20 percent of the population was 4.9 percent in 1944 and 4.7 percent in 1963.

Secretary Wirtz recently summarized these trends. "The confluence of surging population and driving technology is splitting the American labor force into tens of millions of 'have's' and millions of 'have-nots.' In our economy of 69 million jobs, those with wanted skills enjoy opportunity and earning power. But the others face a new and stark problem—exclusion on a permanent basis, both as producers and consumers, from economic life. This division of people threatens to create a human slag heap. We cannot tolerate the development of a separate nation of the poor, the unskilled, the jobless, living within another nation of the well-off, the trained, and the employed.[9]

Given the evidence and these possibilities, the Committee concluded that only changed social norms—public policy on job market restructuring as radical as the impact of cybernation on production—could avert creating a class of citizens who would be alienated from work and income. A first step in developing a solution to meet this problem was seen as the establishment of a general consensus built on the recognition of the dilemmas of income distribution and employment survival in a cybernated society. Accordingly, the Committee indicated that

Cybernation at last forces us to answer the historic questions: What is man's role when he is not dependent upon his own activities for the material basis of his life? What should be the basis for distributing individual access to national resources? Are there other proper claims on goods and services besides a job?[10]

Consequently, it was urged "that society, through its appropriate legal and governmental institutions, undertake an unqualified commitment, to provide every individual and every family with an adequate income as a matter of right."[11]

The Committee reasoned that the question of production efficiency had been solved, and that the central issue was the distribution mechanism. Namely, a cybernated production system must be based on a strikingly different criterion than a manual labor production system. The new system is based on abundance, the old based on scarcity. In was argued that an explicit policy of planned change is needed to facilitate and manage the transition to a cybercultural society.

The program proposed by the Ad Hoc Committee was extraordinarily broad, encompassing provisions dealing directly with the short- and long-term implications of cybernation for the structure of work in society and social equity. At the heart of the recommendations was a call for systematic domestic and international planning which would serve a transformed political economy. The Ad Hoc Committee called for:

1. A massive program to build up our educational system, designed especially with the needs of the chronically undereducated in mind. We estimate that tens of thousands of employment opportunities in such areas as teaching and research and development, particularly for younger people, may be thus created. Federal programs looking to the training of an additional 100,000 teachers annually are needed.
2. Massive public works. The need is to develop and put into effect programs of public works to construct dams, reservoirs, ports, water and air pollution facilities, community recreation facilities. We estimate that for each $1 billion per year spent on public works 150,000 to 200,000 jobs would be created. $2 billion or more a year should be spent in this way, preferably as matching funds aimed at the relief of economically distressed or dislocated areas.[12]

Trade unions were also expected to play a significant role in the transition and beyond by:

1. Use of collective bargaining to negotiate not only for people at work but also for those thrown out of work by technological change.
2. Bargaining for prerequisites such as housing, recreational facilities, and similar programs as they have negotiated health and welfare programs.[13]

Education and public works were seen as major areas of public investment which would allow people to both acquire adequate skills and (particularly in a transition period) have the opportunity to work.

Cybernation was viewed positively rather than negatively, as a key to restructuring work and society rather than simply as a means of enhancing production efficiency within the existing economic system. Thus governmental agencies would recommend ways to stimulate and encourage cybernation through private and public initiatives. But, at the same time, the necessary data would be collected to appraise the social and economic effects of cybernation at different rates of innovation. It was further argued by the committee that high priorities would be placed on efforts

1. To develop ways to smooth the transition from a society in which the norm is full employment within an economic system based on scarcity, to one in which the norm will be either non-employment, in the traditional sense of productive work, or employment on the great variety of socially valuable but ''non-productive'' tasks made possible by an economy of abundance; to bring about the conditions in which men and women no longer needed to produce goods and services may find their way to a variety of self-fulfilling and socially useful occupations.
2. To integrate domestic and international planning. The technological revolution has related virtually every major domestic problem to a world problem. The vast inequities between the industrialized and underdeveloped countries cannot long be sustained.
3. To work toward optimal allocations of human and natural resources in meeting the requirements of society. [14]

Whether or not one considers these proposals appropriate does not diminish the importance of the problems identified by the Committee. A particularly important aspect of the Triple Revolution for this study is its conceptual link of the Black American to the potential damaging impact of cybernetic production techniques.

In carefully reviewing the memorandum, a number of propositions can be identified concerning the impact of cybernation on Blacks. First, the Black American industrial worker (because of his vulnerability to cybernation production technology) will disproportionately suffer immediate unemployment. Second, the occupational mobility ladder up which the American Black's progress was assured in the past will be virtually eliminated because of the impact of cybernation technology. Third, because of the pace of technological change and emerging job skill requirements, the task of reemploying Black industrial workers in other positions will become increasingly more problematic; and, finally, because of the inflexible institutions in the advanced capitalist system, the Black American industrial worker will be made permanently obsolete unless there is a systematic and public intervention to modify the effects of cybernation.

While the Ad Hoc Committee was preparing its position paper on the Triple Revolution in the mid-1960s, Congress was authorizing the formation of the National Commission on Technology, Automation, and Economic Progress to consider a set of issues similar to those raised by the Committee. We now turn to the views of the Commission on cybernation.

The National Commission's Position on the Impact of Automation

The establishment of the National Commission on Technology, Automation, and Economic Progress grew out of a series of bills and public hearings. For example, studies on automation were undertaken in the mid-1950s by the subcommittee on unemployment and the impact of automation. It was stressed at that time that continuing legislative concern was needed if all of the fruits of the potential technology were to be realized while avoiding the individual hardships change might otherwise bring.

President John F. Kennedy (in July, 1963) called for the establishment of a National Commission on Automation. Thereafter, a number of bills were introduced in both the House and Senate. Subsequently, the Ad Hoc Subcommittee was involved in developing a major amendment to the Manpower Act which was accepted by Congress and signed into law by the President on December 19, 1963. It was during the public hearings on the amendment that many of the witnesses called for the establishment of a National Automation Commission. In January of 1964, President Lyndon B. Johnson (in his State of the Union message) proposed the establishment of a Commission on Automation, Technology and Employment. Congressman Holland introduced Bill H.R. 10310 which was designed to implement the investigation of the impact of automation on the work force suggested by Johnson. In April, 1964, public hearings were held and the Commission on Technology, Automation, and Economic Progress was established. While H.R. 10310 established the Commission, H.R. 11611 spelled out its structure and functions.[15]

The 14-man commission was comprised of persons outside the government with established and demonstrated high-level skills and competency in the areas to be investigated. The members were to represent the various segments of industrial and academic life, with at least two members representing labor and two representing management. Appointments to the commission were to be made by the President with the advice and consent of the Senate.

Bill H.R. 11611 directed the commission to:

1. Identify and assess the effects, and the role and pace, of technological change;
2. Identify and describe the impact of technological and economic change on production and employment;
3. Define the areas of unmet community and human needs toward which application of new technologies might most effectively be directed.
4. Assess the most effective means for channeling new technologies in promising directions; and
5. Recommend specific administrative and legislative steps which it believes should be taken by Federal, State, and local governments to support and promote technological change, to continue and adopt measures to facilitate occupational adjustment and geographic mobility, and to share the costs and help prevent and alleviate the adverse impact of change on displaced workers.[16]

Once established, a notion of urgency gripped the Commission as it had the Committee. However, the views of the two groups tended to differ more than concur. The Ad Hoc Committee described the technological advancing period as a new technology—cybernation. The Commission described this technological period by its "elements," namely, the automation of a machine tool, reorganization of an assembly line, substitution of plastics for metals, teaching of a foreign language by electronic machines, etc. So instead of capturing these innovations in a single term that described a common underlying principle (i.e., the self-regulating and self-correcting machine through the principle of feedback) the Commission reasoned in general terms—the pace of technological change.

Regarding its initial impressions of the impact of technology, the Commission started from an optimistic premise. It pointed out that

According to one extreme view, the world—or at least the United States—is on the verge of a glut of productivity sufficient to make our economic institutions and the notion of gainful employment obsolete. We dissent from this view. We believe that the evidence does not support it, and that it diverts attention from the real problems of our country and the world. However, we also dissent from the other extreme view of complacency that denies the existence of serious social and economic problems related to the impact of technological change.[17]

The "extreme" view is presumably the one developed by the Ad Hoc Committee. Since the pace of technological change cannot be measured, the Commission argued, there is no evidence to support the pessimistic view. The Commission concluded: "Our broad conclusion is that the pace of technological change has increased in recent decades and may increase in the future, but a sharp break in the continuity of technical progress has not occurred, nor is it likely to occur in the next decade."[18]

Next, the Commission dealt with the issue of technological change and unemployment. To set the tone for its discussion of this issue, the Commission stressed the need to distinguish the general level of unemployment from the displacement of particular workers at particular times and places.

The main objective of economic policy should be to match increases in productive potential with increases in purchasing power and demand. Otherwise, continued the Commission, the potential created by technical progress will cause waste. Such waste would be manifested in idle production capacity, unemployment, and deprivation.

While rejecting the view that technology would create systematic changes in the relation of labor to production, the Commission believed that some public intervention would be necessary to smooth the transition to a more automated production process. Over the next 10 years, the Commission concluded,

it is at best difficult to separate the technological from other causes of the structural changes that we have been describing. To the displaced employee, or even to the maker of public policy, the precise causes of displacement and unemployment may not even seem important. . . . it is society's responsibility to see that alternative opportunities are available and that blameless individuals do not bear excessive costs.[19]

As for Black Americans, the Commission visualized problems in their adjustment to change:

The adjustment to technological as well as to economic and social change presents special problems for Negroes and other minority groups. No set of measures promoting public and private adjustment will suffice if the avenues to education, jobs, advancement, and the highest achievements of our society can offer are impeded by discrimination.[20]

However, in contrast to the Committee, the Commission placed no special emphasis on the relationship between the jobs of Black Americans and technological advance. The Commission was content to reason that the problems facing the Black American were mainly the result of discrimination or man-made injustices. Thus, the Commission pointed out,

We believe that employers and unions alike have an affirmative duty to make special efforts to aid Negroes and members of other minority groups in obtaining more and better jobs. Such efforts will not in themselves redress the injustices which these disadvantaged citizens have already suffered; but surely they are the very least we should expect from those who profess a belief in democracy.[21]

The Commission's position on the pace of technology change and unemployment, and on the Black American and his adjustment to change, are based on a number of assumptions which reflect an equilibrium model of the economy with self-adjusting capacity, presuming that the government pursues facilitative policies. The Commission argued that, if 1947 is chosen as a dividing point, the trend rate of increase from 1909 to that date was 2 percent per year; from 1947 to 1965 it was 3.2 percent per year. This is a substantial increase, but there has not been and there is no evidence that there will be in the decade ahead an acceleration in technological change more rapid than the growth of demand can offset, given adequate public policies. The Commission also felt that the excessive unemployment following the Korean war, only now beginning to abate, was the result of an economic growth rate too slow to offset the combined impact of productivity increase (measured in output per man-hour) and a growing labor force.

They also argued that technological change, along with other changes, determines who will be displaced. The rate at which output grows in the total economy determines the total level of unemployment and how long those who become unemployed remain unemployed, as well as how difficult it is for new entrants to the labor force to find employment. Unemployment tends to be concentrated among those workers with little education, not primarily because technological developments are changing the nature of jobs, but because the uneducated are at the "back of the line" in the competition for jobs. Education, in part, determines the employability and productivity of the individual, the adaptability of the labor force, the growth and vitality of the economy, and the quality of the society. But we need not await the slow process of education to solve the problem of unemployment.

In its recommendations, the Commission made the following points:

1.   There should be a program of public service employment, providing, in effect, that the Government be an employer of last resort, providing work for the "hard-core unemployed" in useful community enterprises.

2.   Economic security should be guaranteed by a floor under family income. That floor should include both improvements in wage-related benefits and a broader system of income maintenance for those families unable to provide for themselves.

3.   In relation to education, there should be compensatory education for those from disadvantaged environments, improvements in the general quality of education, universal high school education and opportunity for 14 years of free public education, elimination of financial obstacles to higher education, lifetime opportunities for education, training, and unions to provide compensatory opportunities to the victims of past discrimination and stronger enforcement provisions in civil rights legislation relating to employment. Federal, state, and local governments are encouraged to conduct themselves as model employers in the development of new adjustment techniques.

4.   It was also proposed to "humanize" the work environment by adapting work to human needs, increasing the flexibility of the lifespan of work, and eliminating the distinction in the mode of payment between hourly workers and salaried employees.

5.   Finally, a system of social accounts was called for to make possible assessment of the relative costs and benefits of alternative policy decisions along with continuous study of national goals and evaluation of our national performance in relation to such goals.[22]

While there are some similarities in specific recommendations of the Commission and Committee, their basic assumptions and the logic of their analyses are totally divergent. The Commission's position is based on the assumption that an overall framework or unchanging paradigm exists for the whole economy; and that its institutional character is constant (e.g., labor markets, types of labor demanded, types of labor markets, types of labor supplied). The model's purpose is to demonstrate mathematically that all of its components can adjust to levels that are mutually consistent. It assumes, in the long run, that basic model constructs such as consumer preferences, production functions, forms of competition, and factor supply schedules remain unchanged.

The concept of automaticity is an immutable part of the equilibrium model. The system can correct itself if proper economic policies are encouraged. Potential discontinuities caused by either lumpy investments or technological innovations could be evened out over time. Given these theoretical assumptions, the Commission's expectations and policy recommendations are instructed by a quite different causal model than the Committee's (even though both see short-run unemployment resulting from cybernation). In the Commission's world, labor and production will reestablish an equilibrium in the long run. Discrimination or lack of education will be the principal factors if Blacks are unemployed. For the Committee, cybernation has the potential to irreversibly change the labor-production relationship by making human input obsolete and absolutely reducing the amount of work (as we now define it) available. These are critical differences from a policy perspective. It will be argued that the Committee's analyses of cybernation better reflect reality. Thus the optimism of the Commission proved more correct, but only for the short run. Today, technology appears capable of cybernating production to the degree the Committee assumed possible.

Some Evidence in the U.S. and Worldwide of the Current Impact of Cybernation Technology

Because the current cybernation revolution has a consistent internal dynamic, its consequences for the entire industrial world are essentially the same. In America, the nature of work, the distribution of the work force, and the number of industrial workers are changing significantly. In European societies similar changes are occurring. This section will review evidence of the labor-displacing and obsolescing potential of cybernation technology in the United States, with a particular emphasis on the introduction of the robot into the automobile assembly plant and a Black production worker's specific response to this. Further, attention will be given to international evidence regarding the impact of cybernation technology and the consequences facing manual labor.

The Society of Manufacturing Engineers (SME) in 1978 released a survey entitled "Delphi Forecasts Predict Changes." This survey consists of a sampling of predictions made by manufacturing experts. These experts expect that increased application of automation, computer systems, and robotics in the next 10 years will create a greater need for specialized knowledge and skills in both management and worker positions. The survey is divided into two major parts: manufacturing management forecast of events and the year in which they are most likely to occur (the General Forecast), and an assembly technology forecast of events and the year in which they are most likely to occur (the Specific Forecast).

On the basis of the General Forecast, it is clear that new technology will result in irreversible qualitative changes. Particularly noteworthy is the 1988 forecast which states that 80 percent of all in-process and finished inventory will be controlled by a central computer. Also of significance is the 1990 forecast which predicts a 32-hour work week for N.L.R.B. industries. The Specific Forecast is even more revealing of the obsolescing of the general worker. (We shall turn to it later.) The gist of the General Forecast is that the function of manual labor power and/or dexterity will be largely eliminated from the production process. In figure 2-2 the 1982, 1985, 1988, and 1990 forecasts reflect this. Turning to the Specific Forecast in figure 2-3, consider the forecasts for 1985 and 1990. A large percentage of the manufacturing sector is cybernated.

The extent of the introduction of cybernation production techniques into industrial production is also reported from other quarters. It was reported that

> Industrial robots are rapidly moving into the U.S. labor force as manufacturers accelerate automation in order to hold down costs, boost sagging productivity and compete better in world markets. Nearly 5,000 robots currently are toiling away in the U.S., up from 1,300 as recently as 1979. There may be 120,000 robots at work by 1990, predicts Laura Conigliaro, a securities analyst at Bache Halsey Stuart & Shields Inc.[23]

The *Wall Street Journal* indicated that

> Today's robots could replace one million workers by 1990 in the automotive, electrical-equipment, machinery and fabricated-metals industries, concluded a recent study by Carnegie-Mellon University. "Some time after 1990, robot capabilities will be such as to make all (7.9 million) manufacturing operatives (in these industries) replaceable," and three million jobs actually may be lost, says a public-policy professor at Carnegie-Mellon. "Clearly, the members of those unions are at risk."[24]

General Motors, America's leader in auto production, projects the following production job losses by 1990: spot and arc welding, 3–5 percent and 15–20 percent respectively; materials handling, 30–35 percent; paint spraying, 5 percent; assembly, 35–50 percent; and other, 7–10 percent.[25] If we consider simply the SME forecasts and GM's 1990 projections, cybernation will have significant impact on manual laborers. However, the impact is equally significant on the larger scale of the international scene.

The application of cybernation to production is not limited to the United States. The assumption is widely held that a nation's capacity to maintain an eternally viable economy and compete internationally depends upon adopting computer-based production modes. In Britain, for example, this is a recurring theme. A 1978 article noted that

> In the fifties and sixties it was only possible to automate at great expense for one specific task, and hence the product concerned had to have a steady, virtually guaranteed market, engine crank

Figure 2-2. SME's General Forecasts (GF) for Manufacturing Management

The Forecast

Year	The Forecast
1980	The supervision of people will require a different approach in accomplishing objectives as the engineer, technician, and worker of the future demand more responsibility, challenge, and more job enrichment.
	Manufacturing management will require higher levels of specialized technical knowledge than education and experience of a broad generalized nature.
1982	There will be increased influence by manufacturing engineers in the area of government concepts and legislation.
	Better and simpler computer systems will be developed for use in smaller industrial plants.
1985	Production floor workers will be consulted as a matter of course on the structuring of any new jobs or the changing of existing jobs in at least 30% of U.S. industry.
1987	Jobs will be restructured so that in at least 20% of U.S. manufacturing plants each worker (or subgroup of workers) will be doing more work elements or complete operations as against the current practice of single, simple work elements.
	Manufacturing engineering will become a degree program in colleges and universities, offered by at least 33% of the U.S. engineering schools.

Figure 2-2 (continued)

Year	The Forecast
1988	The production control function will be automated to the point that 80% of all in-process and finished inventory is controlled by a central computer. Computers will automatically generate processing plans in 30% of all manufacturing.
1990	The 32 hours/week workweek will become the new standard for organized and N.L.R.B.-affected industries. Fifty percent of the work force on the floor will be highly skilled and trained engineers and technicians will be keeping automated, robotized and computerized plants operating.

Source: The Society of Manufacturing Engineers, Dearborn, MI, 1978

Figure 2-3. SME's Specific Forecasts (SF) for Assembly Technology

Year	The Forecast
1980	There will be a shortage of skilled personnel to service computer-controlled automatic assembly equipment.
1982	Laser and electron beam welding will realize a 50% increase in use. Five percent of the assembly systems will utilize robotic technology. Automatic assembly systems will make more use of automatic diagnostics/fault correction systems.
1983	In structural adhesive applications, solvent systems will be replaced by reactive (100%) solids types even at greater cost under pressure from OSHA and the EPA. Energy shortages will expand the use of room-temperature curing convenience adhesives replacing the majority of heat-curing systems such as epoxies.
1985	Twenty percent of the direct labor in automobile final assembly will be replaced by programmable automation.
1987	Assembly operations will be integrated with other manufacturing operations so that a computer-aided manufacturing system might result.
1988	Fifteen percent of the assembly systems will utilize robotic technology.

Figure 2-3 (continued)

The Forecast

Year	The Forecast
1990	Fifty percent of the direct labor in small component assembly will be replaced by programmable automation.
	The development of sensory techniques will enable robots to approximate human capability in assembly.

Source: The Society of Manufacturing Engineers, Dearborn, MI, 1978

shafts for example. Now, however, with the introduction of microprocessors, automated robots can be used for one task, and then, should demand change, they can be reprogrammed for another; in some cases this takes less than half an hour. Mick McLean of Sussex University's Science Policy Research Unit sees this as the major new development in automation. The process is still expensive, but he is convinced that automation has finally come home to roost because the technology has developed to the point where it is often as flexible as the retraining of people.[26]

As unemployment in Britain began to rise in 1978, the government began to take on a more active role rather than its heretofore advisory role, namely:

the Government made various announcements showing that they had begun to grasp the economic importance and social implications of the increased use of microelectronics. On the economic side, the Department of Industry will provide support for the Industrial users of microprocessors and for the microelectronics industry itself.[27]

Clearly, the British government must play an active role during this cybernation revolution or the future of Britain would be so gloomy as to defy description (in Hines's words).

The British government is attempting to meet the challenge of robotic technology and application. In recent years the British government:

has offered financial incentives to encourage automation. Companies installing robots, for instance, can qualify for government grants totaling as much as 25% of a robotics project. Britain is now funding 32 such projects, and an additional 44 are under consideration. . . . As the British Robot Association's annual conference last year proclaimed, "Failure faces those who don't grasp the opportunities for robotics." Kenneth Baker, British minister of information technology, says the slogan for the 1980's is "Automate or Liquidate."[28]

The Japanese have a relatively long history of adopting industrial robots, and their approach has tended to be more organized and rational. In 1971, it was noted that "The Japanese . . . are busily experimenting with robots as a possible answer to the labor-shortage."[29] Moreover, Japan can make the enviable boast of successfully operating a manless factory. For instance, machines work the night shift unassisted. Only one night watchman patrols the factory floor.[30]

Other countries are also adopting robotic technology. As of 1980, West Germany had 850 production robots, Sweden 600, Italy 500, Poland 360, France 200, Britain 185, Finland 130, and Russia 25.[31]

C. A. Hudson, writing in *Science,* provides an excellent summary of the factors which seem to be common to nations on a worldwide scale in relation to cybernation and production. These are the increasing power and simplification of computers; a widespread appreciation of the practicality of computerized manufacturing and robotic applications; a realization of the impact of computers on people and of people on computers; and a growing awareness of the urgent need for manufacturing innovation.[32]

While much of the emphasis on a global scale is upon the imperatives of cybernation for production efficiency, the potential impact on employment and the work force has not gone unnoticed.

One influential international organization monitoring the effect of cybernation technology is named the Worldwatch Institute. This independent nonprofit research organization was created to identify and focus on "global" problems. Concerning its analysis of the impact of cybernation technology, Worldwatch reported:

> The effects of the microelectronics revolution will be markedly different from those of the industrial revolution, however. The development of industrial technology largely enhanced human physical capabilities, enabling people to harness more energy, process and shape materials more easily, travel faster, and so on. But the development of microelectronics extends mental capabilities, for it increases the ability to process, store, and communicate information, and it enables electronic "intelligence" to be incorporated into a broad range of products and processes.[33]

The Worldwatch report provides a vision of the possible future for blue-collar and white-collar employment:

> The popular science fiction image of the factory of the future pictures an immense computer directing the operation of scores of machines that turn out parts to be assembled by humanoid robots. The reality will not quite match up to this vision, however. Instead of one giant computer, several minicomputers and microprocessors will do most of the machine-minding, and robots on the assembly line will look nothing like humans. And far from being relegated to the distant realm of science fiction, some elements of the factory of the future are already being installed in the industrial plants of today.[34]

On silicon chips and jobs, Worldwatch observed:

> The microelectronic revolution could affect employment in enterprises ranging from steel works to banks; no technology in history has had such a broad range of potential applications in the workplace . . . products incorporating microelectronic devices generally require significantly less labor to produce than the goods they replace, a fact that extends the employment implications of technology well beyond its direct impacts on automation.[35]

Finally, Worldwatch raised the issue that

> It is easy to point to the advantages of raising productivity with new technologies, but if those advantages are won at the expense of displaced workers, the fruits of technological change will be bitter indeed. And, as already outlined, the pervasive nature of microelectronics, coupled with its potential for introducing labor-saving change in both blue-collar and white-collar industries, will heighten the problem of adjustment.[36]

Worldwatch is clear on the issues of microelectronics, robotics, silicon chips and labor displacement—the advanced cybernetic techniques will heighten both the problem of massive unemployment and adjustments. Indeed, the forebodings of the Worldwatch Institute are similar to those of the Ad Hoc Committee, with apparently a better evidential base. There is no doubt that cybernation is occurring on a world scale and the potential for seriously changing the relationship of human labor and production exists in a form that is qualitatively different from the disequilibrium states in the past. Now the possibility exists that production levels adequate for society can occur without requiring anywhere near full participation in the labor force. At the same time, the implications of cybernation do not automatically become widely recognized topics of public discussion in the dialogue of a nation or even among those most directly involved. This is certainly true in the United States at the present time, even though the process of cybernation is occurring in automobile production with large numbers of Black workers at risk. And this specific case provides a means to examine a number of empirical and policy dimensions of the relationship of cybernation and employment and provides the focus for the remainder of this study.

3

The American Auto Industry and the Imperative to Shift from Human Labor to Robotics

Cybernation technology is achieved through microprocessors, analog computers, digital computers, numerical control, and industrial robotry. Two basic applications are computer-aided manufacturing (CAM) and computer-aided design (CAD), which are rapidly coming to be viewed as the foundation of future industrial production. The process industries have successfully shifted to reliance on the analog computer. Manufacturing industries are developing an increased reliance on digital computers, numerical control, and robotry. The American automobile industry represents one of the best current examples of such a shift from reliance on manual power and/or dexterity to cybernation technology—robotic production.

The factors driving the auto industry's shift to this form of technology are several. One is the industry's sagging productivity and profits (in spite of the fact it has the most highly paid industrial workers). High labor costs have made the possibility of shifting to the use of industrial robots particularly attractive. The critical importance of the auto industry to the American economy has put great pressure on the government to support management decisions to utilize cybernation technology. This chapter will consider the conditions that have lead to the introduction of production robots by American automobile manufacturers, and to the expectation that they will be widely used in the future.

The Significance of the Auto Industry to the U.S. Economy

One question to be answered in this chapter is how important the auto industry is to the U.S. economy. In a 1980 Congressional hearing, Senator Reigle described the auto industry as:

> the keystone of this nation's economy. It directly creates one of every 12 manufacturing jobs and generates prime demand for such basic industries as steel, aluminum, rubber, textiles, machine tooling and increasingly, electronics. It affects the economy of every state, and its health is vital to some 50,000 small and medium-sized supplier firms across the 50 states, and to some 27,000 auto dealers.

However, he also noted that "The U.S. auto industry today is in serious difficulty. Sales and production have fallen by the largest amount in 20 years. Last month, 295,000 auto workers were out of work, causing layoffs of an additional 885,000 workers in related industries, a total well above one million workers."[1]

The central role of the auto industry in the economy is easily referenced. As table 3-1 indicates, 77 percent of all U.S. production workers are employed in automotive-related industries. Of equal importance is the fact that this industry is also a major consumer of raw materials, and it generates other related industrial activities (see tables 3-2 and 3-3). The wide geographic distribution of the auto industry is indicated in figure 3-1. Facilities are located in over 30 states. Obviously, the ability of the industry to operate effectively and to compete is critical ·to the American economy.

Table 3-1. Production Workers as 77% of Vehicle and
Equipment Employees

MOTOR VEHICLE AND EQUIPMENT MANUFACTURING
EMPLOYMENT (Annual Average)

Year	All Employees (000)	Production Workers		
		Number (000)	Percent of Total Employees	Average Hourly Earnings
1979	982.8	759.7	77.3%	$9.07
1978	997.2	776.0	77.8	8.51
1977	942.0	730.5	77.5	7.87
1976	881.0	682.4	77.5	7.09
1975	792.4	602.4	76.0	6.44
1974	907.7	687.5	75.7	5.87
1973	976.5	754.9	77.3	5.46
1972	874.8	676.0	77.3	5.13
1971	848.5	655.4	77.2	4.72
1970	799.0	605.3	75.8	4.22
1969	911.4	708.0	77.7	4.10
1968	873.7	680.8	77.9	3.89
1967	815.8	626.9	76.8	3.55
1966	861.6	670.3	77.8	3.44
1965	842.7	658.9	78.2	3.34

Note: These figures are for the Motor Vehicles and Equipment Manufacturing Industry (SIC 371) as defined by the Standard Industrial Classification system. Many others are employed in manufacturing automotive components who are classified in other industries.

Source: U.S. Bureau of Labor Statistics

U.S. Senate, Joint Hearing before the Subcommittee on International Finance and the Subcommittee on the Economic Stabilization of the Committee on Banking, Housing, and Urban Affairs, *The Automobile Industry and World Economy,* 96th Cong., 2nd sess., June 18, 1980, 77.

Table 3-2. The Auto Industry: A Major Customer for Raw Materials

Material	U.S. Total Consumption	Automotive Consumption	Automotive Percentage
MALLEABLE IRON (000 of Tons)			
1978	816	435	53.3
1977	826	445	53.9
1976	846	470	55.6
1975	731	344	47.1
1974	914	417	45.6
NATURAL RUBBER (Metric Tons) (2)			
1978*	764,654	590,577	77.2
1977*	780,000	623,000	79.9
1976*	730,727	491,000	67.2
1975*	669,966	497,370	74.2
1974	707,722	552,000	78.0
RECLAIMED RUBBER (Metric Tons) (2)			
1978*	118,732	73,415	61.8
1977	111,000	80,000	72.1
1976*	81,892	N.A.	N.A.
1975*	100,216	N.A.	N.A.
1974	142,292	N.A.	N.A.
SYNTHETIC RUBBER (Metric Tons) (2)			
1978*	2,436,399	1,377,592	56.5
1977*	2,464,000	1,464,000	59.4
1976*	2,175,255	1,272,000	56.2
1975*	2,022,431	1,196,189	59.1
1974	2,351,243	1,383,000	58.8
ZINC (000 Tons)			
1978	1,127	337	29.9
1977	1,103	332	30.0
1976	1,140	380	33.3
1975	925	308	33.3
1974	1,294	431	33.3

Material	U.S. Total Consumption	Automotive Consumption	Automotive Percentage
ALLOY STEEL (000 Tons) (1)			
1978	10,557	2,444	23.2
1977	8,760	2,062	23.5
1976	8,108	1,859	22.9
1975	8,436	1,314	15.6
1974	10,179	1,803	17.7
STAINLESS STEEL (000 Tons) (1)			
1978	1,191	197	16.5
1977	1,118	190	17.0
1976	1,019	187	18.4
1975	757	119	15.7
1974	1,345	165	12.3
TOTAL STEEL (000 Tons) (1)			
1978	97,935	21,254	21.7
1977	91,147	21,492	23.6
1976	89,447	21,351	23.9
1975	79,957	15,214	19.0
1974	109,472	18,928	17.3
ALUMINUM (Millions of Pounds)			
1978	14,390	2,374	16.5
1977	12,964	2,270	17.5
1976	12,747	1,894	14.9
1975	9,929	1,167	11.8
1974	13,732	1,792	13.0
COPPER AND COPPER ALLOYS (Millions of Pounds)			
1978	6,678	801	12.0
1977	6,206	783	12.6
1976	5,736	718	12.5
1975	4,704	507	10.8
1974	6,293	600	9.5
COTTON (480 lb Bales)			
1978	6,357,000	52,530	0.8
1977	6,633,000	66,330	1.0
1976	6,561,130	83,920	1.3
1975	6,350,690	73,460	1.2
1974	6,533,460	87,090	1.3

*In 1975 the unit of measurement changed from "long tons" to "metric tons." To convert "long tons" to "metric tons" multiply "long tons" by 1.016.

(1) "Automotive Consumption" of steel may be understated as shipments to the automotive market from steel centers and distributors are not included.

(2) Includes all rubber products classified as "Transportation products" but excludes "Mechanical Rubber Goods," i.e., rubber weather stripping, grommets, motor mounts used in automobiles.

Note: For most materials listed the automotive consumption includes materials used for cars, trucks, buses and replacement parts.

Source: Compiled by Motor Vehicle Manufacturers Association of the U.S., Inc., from various trade sources.

U.S. Senate, Joint Hearing before the Subcommittee on International Finance and the Subcommittee on Economic Stabilization of the Committee on Banking, Housing, and Urban Affairs, *The Automobile Industry and World Economy*, 96th Cong., 2nd sess., June 18, 1980, 79.

Table 3-3. Other Industries Generated by the Automobile Industry

Industry Producing	Automotive Shipments (Millions)	Estimated Automotive Employment (000)
Narrow Fabric Mills	$ 32.9	1.1
Tufted Carpets and Rugs	185.4	2.2
Paddings and Upholstery Filling	70.5	1.4
Tire Cord and Fabric	949.4	10.4
Automotive and Apparel Trimmings	1.667.6	24.2
Fabricated Textile Products, Nec	385.2	10.1
Public Building & Related Furniture	166.2	4.6
Die-Cut Paper and Board	21.8	.3
Paints and Allied Products	859.6	8.6
Chemical Preparations, Nec	427.1	3.7
Tires and Inner Tubes	7.336.9	103.0
Rubber and Plastics Hose and Belting	100.0	2.1
Fabricated Rubber Products, Nec	1.087.0	27.4
Miscellaneous Plastic Products	2.430.6	49.3
Flat Glass	717.9	8.6
Products of Purchased Glass	89.8	2.0
Asbestos Products	271.5	3.9
Gray Iron Foundries	3.191.0	56.9
Hardware, Nec	2.203.1	43.1
Bolts, Nuts, Rivets and Washers	179.5	3.5
Automotive Stampings	9.470.3	131.9
Steel Springs, except Wire	370.6	5.4
Internal Combustion Engines, Nec	1.322.2	15.6
Metal Working Machinery, Nec	188.0	4.1
Refrigeration and Heating Equipment	1.562.9	25.7
Carburetors, Pistons, Rings, Valves	1.082.8	26.6
Motors and Generators	458.9	91.2
Electric Lamps	223.1	3.9
Vehicular Lighting Equipment	663.5	14.6
Radio and TV Receiving Sets	622.1	9.7
Storage Batteries	1.458.0	19.8
Engine Electrical Equipment	2.796.8	59.7
Total	**$42,592.2**	**774.6**

Nec—Not elsewhere classified.

Note: The above tabulation is incomplete since it contains only those industries which specifically identify automotive parts data in the *1977 Census of Manufactures*.

Automotive employment is estimated by the Motor Vehicle Manufacturers Association by assuming that employment is in direct proportion to the ratio of automotive shipments to total shipments of the industry.

Source: Compiled by the Motor Vehicle Manufacturers Association of the U.S., Inc. from U.S. Bureau of the Census, *1977 Census of Manufactures*.

U.S. Senate, Joint Hearing before the Subcommittee on International Finance and the Subcommittee on Economic Stabilization of the Committee on Banking, Housing, and Urban Affairs, *The Automobile Industry and World Economy*, 96th Cong., 2nd sess., June 18, 1980, 77.

Figure 3-1. National Distribution of the U.S. Automobile Industry

▲ ASSEMBLY PLANTS   There are 96 Assembly Plants located in 85 Cities in 31 States
△ PARTS PLANTS   There are 248 Parts Plants located in 158 Cities in 27 States
● PARTS DEPOTS   There are 205 Parts Depots located in 133 Cities in 33 States
■ PROVING GROUNDS   There are 34 Proving Grounds located at 37 Centers in 12 States
                                 Independent Supplier Firms are in hundreds of other cities.

Source: "MVMA Motor Vehicle Facts and Figures '80," Motor Vehicle Manufacturers Association of the United States, Inc., Detroit, Michigan, p. 16.

The Basis of the Imperative to Shift to Cybernetic Production Methods

The performance of the auto industry has, in fact, been a matter of considerable concern in both the private and public sectors for more than a decade, especially since at least one major company has come close to bankruptcy. As figure 3-2 shows, the earnings of the Big Four (American Motors, Chrysler, Ford, and General Motors) have been relatively flat—after adjusting for inflation—over the past 10 years. While the automotive industry's earnings were up only 14 percent during this period, the earnings of the Fortune 500 firms doubled. Figure 3-3 shows the worldwide earnings of the industry between 1960 and 1977. The top line is not adjusted for inflation. In constant 1960 dollars, the industry's return on sales dropped from a high of 7.8 percent in the 1963–65 period to 3.8 percent in 1975–77. These data indicate the industry's inability to sustain its historical profitability.

Figure 3-5 shows that in the early sixties returns on sales for the automotive industry were a bit higher than the average for the Fortune 500, while profit margins for both declined. Since then, the automotive decline has been sharper, and it represented a lower ratio by 1977. Each of the major U.S. automotive manufacturers has experienced declining returns on sales since the early sixties (see figure 3-4), with those of Chrysler and AMC being lower than those of GM and Ford. AMC has incurred a substantial and cumulative loss since 1963, while Chrysler suffered from 1974 to the first quarter of 1978 average profits of only $36 million a year.

Figure 3-2.   Trend of Automotive Industry Earnings Compared with
Fortune ''500'' (1960 Dollars)

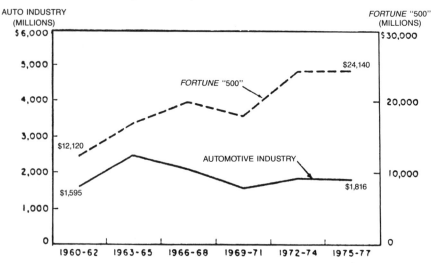

Figure 3-3. Automotive Industry Earnings

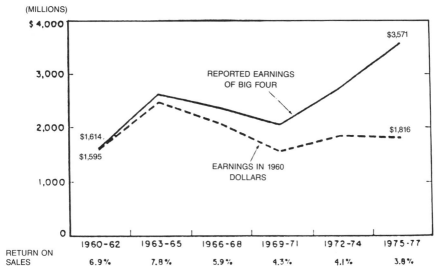

Figure 3-4. Automotive Big-Four Returns on Sales

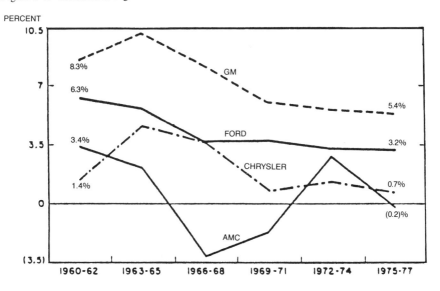

Figure 3-5.   Automotive Industry Returns on Sales Compared with Fortune "500"

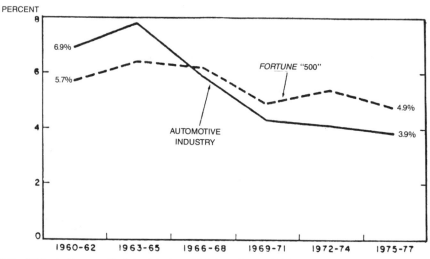

Source: U.S. Senate, *Government Regulation of the Automobile Industry, Hearings,* 96th Cong., April 26, 1979, 354–55.

That the American auto industry was in deep trouble became even clearer after it was noted that its market share (domestic) had declined tremendously during a five-year period. Figure 3-6 shows that from 1974 to 1979, GM's market share increased from 41.8 percent to 47.7 percent, while Ford's decreased from 25 percent to 20 percent. Chrysler's performance also was poor: its market share fell from 13.6 percent to 9.3 percent.

The effect of this decline in market share is ultimately reflected in falling profits of the Big Three. In table 3-4, GM's historical profit performance is consistent with the growth of its domestic market share. Ford and Chrysler for the years 1974 and 1979 both experienced declines in profits which are consistent with their loss of domestic markets.

In discussions of the causes of and possible solutions for the decline of the American auto industry, two factors have received considerable attention: labor costs and the substitutability of robotic production for human labor. For the Japanese, the application of cybernetic techniques to their auto industry has resulted in a virtual robotized production line. This has been seen as a major factor in their competitive advantage over American auto makers. Adopting this strategy is viewed by many as a necessary step in revitalizing the industry in the United States. As it was put at a recent Senate hearing: "A Datsun plant in Japan manned by 67 workers produces 1,300 cars per day. The United States cannot meet this competition without moving substantially in the same direction, more efficient production of higher quality, more fuel efficient automobiles."[2]

The Japanese auto manufacturers, using robotic technology, have gained a

Figure 3-6. Market Shares of the American Automotive Market, 1974–79

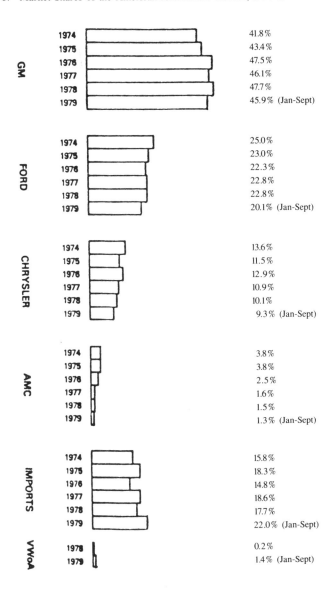

GM
1974   41.8%
1975   43.4%
1976   47.5%
1977   46.1%
1978   47.7%
1979   45.9% (Jan-Sept)

FORD
1974   25.0%
1975   23.0%
1976   22.3%
1977   22.8%
1978   22.8%
1979   20.1% (Jan-Sept)

CHRYSLER
1974   13.6%
1975   11.5%
1976   12.9%
1977   10.9%
1978   10.1%
1979   9.3% (Jan-Sept)

AMC
1974   3.8%
1975   3.8%
1976   2.5%
1977   1.6%
1978   1.5%
1979   1.3% (Jan-Sept)

IMPORTS
1974   15.8%
1975   18.3%
1976   14.8%
1977   18.6%
1978   17.7%
1979   22.0% (Jan-Sept)

VWoA
1978   0.2%
1979   1.4% (Jan-Sept)

Source: U.S. House of Representatives, *The Chrysler Corporation Financial Situation, Hearings,* 96th Cong., 1st Sess., Oct. 1979, 476.

Table 3-4.   Big-Three Historical Profits

(Dollars in millions)

Year	GM	FORD	CHRYSLER
1979 (6 mos)	$2,445.2	$1,107.2	$(273.5)
1978	3,507.8	1,588.9	(205)
1977	3,337.5	1,672.8	163
1976	2,902.8	983.1	423
1975	1,151.1	322.7	(260)
1974	950.1	360.9	(52)
1973	2,398.1	906.5	255
1972	2,162.7	870.0	220
1971	1,935.7	656.7	84
1970	609.1*	515.7	(8)
1969	1,710.0	546.5	99
1968	1,730.0	626.6	290.7
1967	1,627.3	84.0	200.4
1966	1,793.4	621.0	189.2
1965	2,125.6	703.0	233.4
1964	1,734.8	505.6	213.7

*UAW Strike

Note: Parentheses indicate a loss for that year.

Source: U.S. House of Representatives, *The Chrysler Corporation Financial Situation, Hearings,* 96th Cong., 1st sess., Oct. 1979, 488.

decided productivity/labor-saving edge over their American counterparts. But the robotic production techniques are only part of the story. The Japanese produced a small, fuel efficient car that fulfilled an American consumer demand, particularly following the 1973 OPEC oil embargo. Robotic production allowed plants to meet this market demand while boosting overall production efficiency (in terms of higher quality) and lower general labor costs. The *New York Times* reported that a major Japanese auto company has a welding line that consists of 18 separate work processes, each of which takes only three minutes. This work is done by 80 welding robots and, more importantly, they are ready to handle additional workloads if necessary.[3]

As the *Times* further reported, Japan (at the end of 1981) had close to twice as many programmable robots in operation (14,000) as the rest of the world combined. The United States was a distant second (see figure 3-7).

Figure 3-7. The Japanese Quest to Robotize Their Production Lines

The Robot Population

Programmable Robots In Operation	
JAPAN	14,000
UNITED STATES	4,100
WEST GERMANY	2,300
FRANCE	1,000
SWEDEN	600
BRITAIN	500

Source: Daiwa Securities America; Robot Industry Association; Wertheim & Co.

How Robots Are Used
Breakdown of total robot usage by function

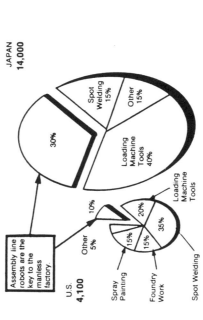

JAPAN
14,000

Spot Welding 15%

Other 15%

Loading Machine Tools 40%

30%

Assembly line robots are the key to the manless factory.

U.S.
4,100

Other 5%

10%

Spray Painting 15%

Foundry Work 15%

15%

20%

Loading Machine Tools

35%

Spot Welding

Source: Wertheim & Co.

Source: *New York Times Business*, Sunday, December 13, 1981, Section 3, p. 1.

The inherent attractiveness of cybernated production assumed an imperative nature for the American auto industry when the cost of human labor, which was growing exponentially, was factored into production costs. Between 1957 and 1979, labor costs (which included fringe benefits and which were adjusted for inflation) spiralled from three dollars per hour to eleven dollars per hour, while return on sales and profits declined. These labor cost data are in figure 3-8. However, when comparing variable costs as a percentage of sales, the argument that reductions in labor costs were necessary is easily reinforced (see table 3-5). This view is forcefully stated in the trade journal *Machinery and Production Engineering*:

> the payment of high wages to workers who cannot be described by any standards as anything more elevated than machine minders is rapidly becoming unattractive, and where a man is employed solely for unloading one machine and loading another (as is so often the case today) the substitution of a robot is not only a glaringly obvious course but also increasingly easy to justify financially. Moreover a robot is not subject to the random variations in performance in a human being, and is for all practical purposes working as hard, as conscientiously, and as consistently at the end of the shift as it is at the beginning.[4]

To be sure, the evidence of the effectiveness of shifting to robotic production was based on the Japanese experience. Because, in part, of Japan's success in automobile production, American auto manufacturers were encouraged to introduce robotic production methods. These are particularly attractive as a means of reducing labor cost; and robotic production has become the centerpiece in the recovery strategy of the American automobile industry.

Some General Applications of Robots to Production-Line Activities

To better appreciate the overall significance of robotic technology, two brief discussions are useful. The first concerns the growing significance of robotic technology, and the second involves the production-line applications of robotics.

Table 3-5. Estimated Variable Costs[a] (as a Percentage of Sales)

Firms	1974	1975	1976	1977	1978
GM	71.5	72.4	69.6	69.2	69.6
Ford	80.0	81.1	78.0	77.0	78.6
Chrysler	86.6	88.9	83.8	85.8	86.6

[a]Variable costs are costs that change directly with production volume. Examples are: labor, direct materials, parts purchased from outside suppliers, and electric power.

Source: U.S. House of Representatives, *The Chrysler Corporation Financial Situation, Hearings,* 96th Cong., 1st sess., Oct. 1979, 550.

Figure 3-8. History of Labor Cost in the U.S. Automotive Industry

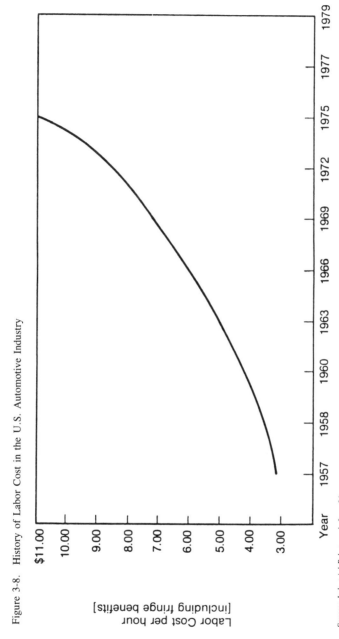

Source: *Industrial Robots*, vol. I, p. 36.

How important are robots becoming to society? To answer this question, consider the comments of one of America's leading robotic experts:

> Robots are beginning to emerge as economic realities. Robotics is fast becoming an international race. The new industry promises a long range solution to the problem of reaching competitive productivity rates. . . . Strategic use of robots can be a significant factor in reaching the 5 to 10 percent rates of other countries. The race may start in those sectors of the U.S. that are experiencing an erosion of their industrial base. . . . Our present machine systems are in the final stages of their life cycle. The older automation age is dying and the cybernetic machine age is starting its life cycle. The machine system will be replaced by a "machine and machine system" or a "cybernetic productivity system." Machines such as blue collar robots will be commanding and controlling other machines and will be working with a new generation of production workers in cybernetic productivity systems.[5]

The *Wall Street Journal* gives three very good examples of advancements made in cybernetic machines:

> Thanks to the capacity of today's computers and microprocessors and the ingenuity of the engineers who design them, the human instinct to communicate orally with inanimate machines isn't fruitless any longer:
> —At the Ford Motor Co.'s parts warehouse in Teterboro, N.J., an employee picks up a package at the loading dock, reads the destination into a microphone and puts the package on a conveyor belt. Without further ado, conveyor belts carry it to the correct storage shelf.
> —Meat inspectors at six Wilson Foods Corp. packing plants weigh and grade hog carcasses as soon as the animals are slaughtered and dictate the information orally into a computer. It prints up a record of each farmer's hogs and later writes a check for the farmer. The process eliminates blood-stained paper work and the inaccuracies of typing into a terminal.
> —State employees in Illinois make calls on the state's private telephone line even when they are outside the office by calling a computer, reciting an authorization number and getting oral approval from the computer. It then asks for the phone number and dials it, just as human long-distance operators do.[6]

With the growth and development in robotic technology, engineers in this field foresee wider ranges of production-line activities. Consider some recent observations made by Paul Kinnucan regarding advances made in smart robotic technology:

> Equipped with television cameras for eyes and microcomputers for brains, these "smart" robots are gradually acquiring the ability to see, touch, hear and make decisions much like the human workers they are intended to replace. Indeed, observers expect the coming decade to launch the "factory of the future" where almost all manufacturing will be done by robots assisted by computer-controlled machine tools and transfer lines.
> . . . . Such capabilities will enable robots to do factory jobs now requiring human labor or costly specialized machinery. Also under development are mobile smart robots that could one day scavenge the seabed for minerals or sunken ships or descend mine shafts to dig coal.
> Sea or coal mining by mobile robots could be decades away. Improvements in robots' sensory and manipulative capabilities, however, should allow them to play a greatly expanded role in the factory by the end of this decade.[7]

In terms of specific applications to the auto industry, General Motors has been heavily involved in research and development in robotic technology. It has produced a programmable universal machine for assembly (PUMA).[8] Figure 3-9 contains a diagram of the PUMA system. It is designed to use robotics interchangeably with humans in the same work place, working at the same speeds.[9] The system can take the form of straight-line indexing, rotary indexing, or any possible configuration that is applicable to a given product or location.

The current applications of cybernated techniques in auto production are diverse, and they are increasing. Consequently, as figure 3-10 indicates, the number of production-sector functions heretofore performed by humans but in which human labor is now obsolete is diverse and increasing. We will now review the role of cybernation in a basic set of auto production-line activities which commonly have provided jobs for Black workers. These include material handling, die casting, forging and heat treating, foundry, welding, finishing, and assembly.[10]

Material Handling

Tooling to allow industrial robots to perform as material handlers can be conveniently divided into two main categories: (1) mechanical gripper, and (2) surface-lift devices. Thus, mechanical grippers for material handling typically employ movable fingerlike levers paired to work in opposition to each other (a direct replication of the human hand). A single mechanical hand, for example, might have one or several sets of opposed fingers. A robot can have one or several hands. Surface-lift devices are exemplified by vacuum pick-ups and electromagnets. Vacuum pick-ups handle durable or delicate nonferrous materials with flat or gently curved surfaces. Mechanical material handling grippers are industrial robots that transfer automotive engine heads from a power-rolled conveyor to a monorail conveyor.

Vacuum pick-up hand tooling is an adaptation of the simple vacuum cup. For instance, consider an example where sheets of glass are handled for an edge-grinding operation. The robot loads and unloads two edge grinders alternately, picks up unground glass sheets on a stacker. Inherent in these robots are control techniques—the more sophisticated the control technique, the more efficient the robot as a reliable factor of production. The control aspect system can be point-to-point or continuous path.

Die Casting

According to one authority:

Earliest efforts to achieve operator-free automatic cycling of die casting machines started many years ago and achieved a high degree of efficiency in the production of relatively small, simple

Figure 3-9. GM's PUMA Concept, Using Robots and Human Beings

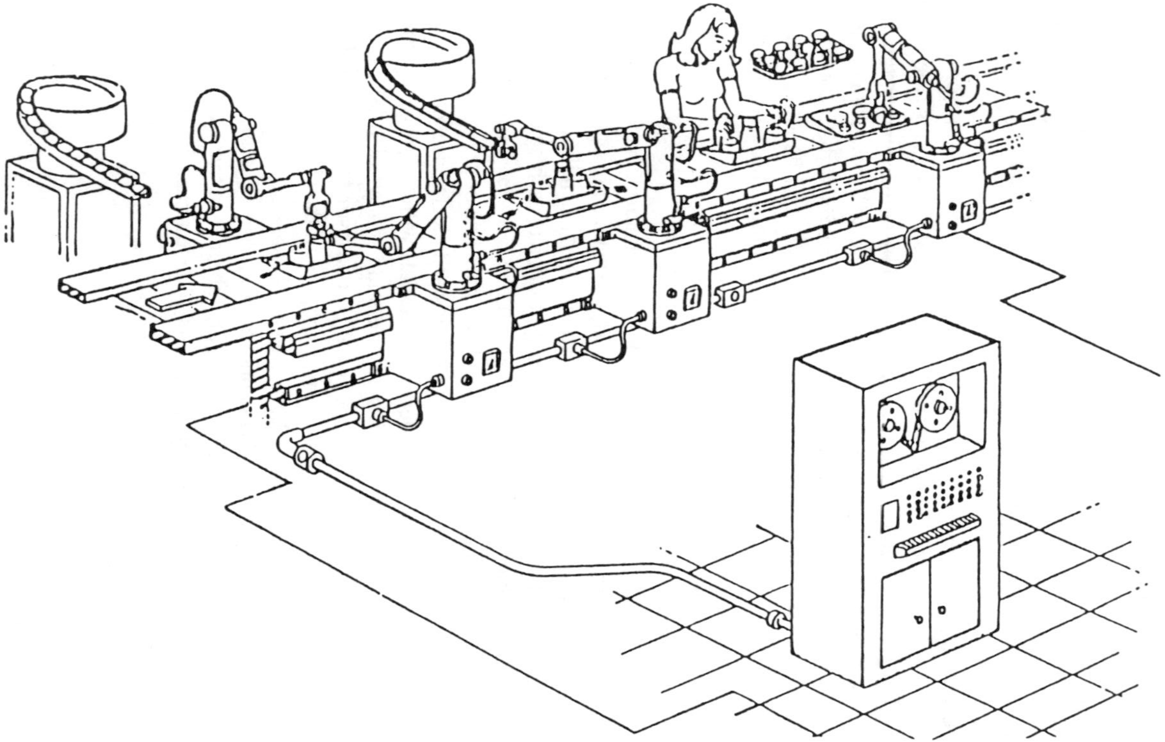

Source: *Iron Age* (Nov. 28, 1977), p. 35.

Figure 3-10.   A Sample of the Jobs Robots Can Do

Die Casting
    Unload 1 or 2 DCM—quench—trim—die care—insert loading—palletizing

Forging
    Drop forge—upsetter—roll forge—presses

Stamping Presses
    Load/unload—press-to-press transfer

Welding
    Spot welding—press welding—arc welding

Injection & Compression Molding
    Unload 1 or 2 IMM—trim—insert loading—palletizing—packaging

Investment Casting
    Wax tree processing—dipping—manipulating—transferring

Machine Tool Operations
    Load—unload—palletizing—machining center—machine-to-machine transfer—operating
    with lathes, chuckers, drilling machines, broaches, grinders, multi-turret machines, etc.

Spray Coat Application
    Mold releases—undercoats—finish coats—highlighting—frit application—sealants

Material Transfer
    Automotive assembly—automotive parts—glass—textile—ordnance—appliance manufacturing
    —molded products—heat treating—paper products—plating—conveyor and monorail loading
    and unloading—palletizing and depalletizing.

Source: *Industrial Robots,* vol. 1, p. 36.

parts, cast from the lower melting temperature alloys on small machines. . . . In the sixties, most early robot applications in die casting were simply to replace casting machine manual operation. . . . Although this simple substitution of robot for manual operator was intended to merely assure production, it was also recognized that the manual operator's function involved far more than just extracting castings from the machine.[11]

The Unimate—the 2000 series and 4000 series—contains both five and six axes. This means that the manual laborer's mechanical functions can be performed by a programmable robot with six axes of freedom and/or continuous path capability.

The following factors have encouraged the substitution of manual labor for the industrial robot in die casting: (1) the escalation in the price of zinc, which has forced a shift to the use of aluminum (which in turn has accelerated a trend toward automatic lading of metal in cold chamber units and an increase in the tonnage capacity of the new machines delivered); and (2) robotic strength and linear stroke to manipulate large castings at high speed. It is reported, for example, that "A major benefit was realized when a large die casting facility with approximately 50% of its die centers automated with 40% Unimate industrial robots was able to maintain just under 40% production during a strike."[12]

Forging and Heating

Various forging machines have been utilized with industrial robots for several years. One can find upsetters, roll forges, programmable hammers, impacters, and forge presses all interfaced with robots in single die or three-die configurations. Forge die and hot trim die operations have been installed. One industrial user of robots made the following observations:

> Cost reduction programs are a continuing effort to hold prices while maintaining margins. Of the many potential cost savings associated with particular products, material handling improvements fall into two basic categories: hardware which assists the worker and makes him more effective, and hardware which replaces man. Both hard automation and programmable automation must be considered. [13]

As far as hard automation is concerned (and particularly this case of aluminum forging), the jobbing-shop nature of the aluminum industry prevented successful hard automation. But this industry has, only recently, successfully replaced man with programmable automation.

A number of innovations in auto-related forging and heating involving the replacement of manual labor are cited in "Unimation Application Notes." Consider industrial forging: "A mid-west auto manufacturer is using a 2000A Unimate to feed billets through a two-cavity die forging press where they are forged into raw differential side gears." [14] Historically, the Black production worker (unskilled) performed this activity in the automotive industry. The innovative Unimate 2000A thus displaces one operator per shift in a three-shift operation. [15]

In another example, "A major automotive supplier has produced a 4000Z Unimate to be used in the production of forged diesel engine crankshafts." [16] This innovation—a very high production operational innovation—displaces two and two-thirds operators on a two-shift basis. One final example is noteworthy. Regarding forging and trimming, Chambersburg Engineering Company has developed an impact forging system incorporating a model 2000B Unimate industrial robot. The justification advanced for this innovation—a completely automated system—is as follows: "Automating this operation with one Unimate robot has put production at a level of over that of five manual operators. In addition, hot trimming has measurably enhanced productivity by increasing the integrity of the grain flow of the part, resulting in higher part quality." [17]

Foundry

The industrial robot and its flexibility provides efficient use in large shops with long production runs and small independent foundries where lot sizes of 1000 castings or less are common. That is to say, lot sizes usually range from a few kilograms to over 50 kilograms. The task assigned to the robot in the foundry where Black workers were usually found is to successively dip delicate wax molds (trees) into a series of slurries and fine sands to build up a shell which eventually becomes a casting mold. That this innovation in the foundry increases the level of productivity is clear from the following. Human manipulation was found to be at best inconsistent. The quality of the mold and the finished part is not known until the casting is made. But, by this time a large investment in time and material has been made in producing what is all too often referred to as scrap. In some cases the scrap rates have been reduced from 85 percent to under 5 percent.[18]

> The entire process is under direct computer control. Process parameters such as slurry temperatures and viscosities, drying temperatures, etc. are monitored and adjusted. In addition, robot programs are stored in the disc file, ready to be dumped into the robot's local memory as soon as the wax tree identification (and hence correct robot program) is made.[19]

Welding

General purpose industrial robots can maneuver and operate a spotwelding gun to place a series of spot welds on flat, simple-curved, or compound-curved surfaces. In production-line operations on auto bodies, stop-and-go rather than continuous line motion is preferred. Ironically, the much beleaguered Chrysler Corporation was a pioneer in the instruction of the welding robot. For instance:

> In 1971, Chrysler installed several robots for evaluation in parts welding operations. Now fully proven and operational, two robots are spotwelding various assemblies used in the company's "Lance" model automobile. . . . In another operation at the Hamtramck plant, a robot, assisted by one operator, makes 48 separate movements to perform 22 welds in 74 seconds on lower deck panel assemblies. Previously, two welders were required to attain the same rate of production. . . . In all welding operations where robots are used, reliability is impressive. Operating two 8-hour shifts per day, the robots require maintenance about once every 28 working days. Chrysler industrial economists estimate that the robot installation will amortize itself in less than two years.[20]

A much more recent example of Chrysler fully employing robotic welding is found in its Newark, Delaware assembly plant. This plant builds the "K" car; and it has been redesigned internally to accommodate the more efficient robot. As a comparison, the old welding system had welders with cumbersome water-cooled suspended spotwelding machines placing welds in up to 500 spots around each body as it moved on the build truck down the towline path. The increased speed of the line made the job tedious and unpleasant. But after the plant was redesigned, productivity increased substantially. The new system has preprogrammed controls and it interfaces with different kinds of conveying, lifting, and positioning equipment.

Figure 3-11 is a sketch of the automatic welding line introduced in the Newark plant. The scissor lifts at each end in the illustration are to transfer autos to the

Figure 3-11.   A Sketch of Automatic Welding at the Chrysler Assembly Plant in Newark, Delaware

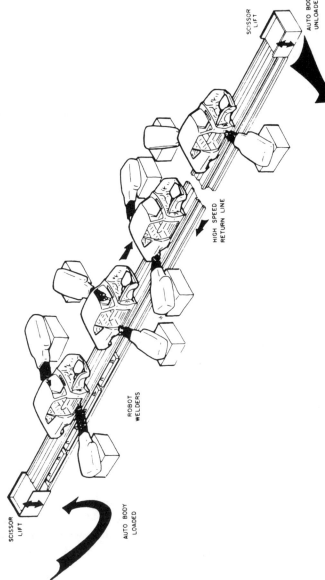

SCISSOR
LIFT

AUTO BODY
LOADED

ROBOT
WELDERS

HIGH SPEED
RETURN LINE

SCISSOR
LIFT

AUTO BODY
UNLOADED

Source: *Industrial Robots*, vol. II, p. 155.

high-speed return line. Because of the responsibility associated with this innovation it can be viewed as operating like a factory within a factory rather than like a segment of a production line. The area that this line encompasses contains more than a dozen programmable controllers (PC). Moreover, there is one PC on each of the 13 robot welders, plus the master PC which controls the stackers and the shuttle car line. There is also another PC employed for testing and standby uses. As one report puts it: "The Newark plant points out how challenges to continuous handling and productivity are being met with good hardware and software on the one hand and user savvy on the other."[21]

Paint Finishing

This section discusses finishing in conjunction with painting. In his article, "Robots in Paint Finishing," Norman N. Fender noted that

> robots answer a long awaited need in the finishing industry. No longer is it necessary to subject man to dangerous and unhealthful environments which are typical of many spray finishing operations. Finally machines are available that can duplicate human sprayers' movements and limited decision making processes.[22]

Echoing this same enthusiasm for robotization in finishing, Jack Shaneyfelt noted that

> as environmental and health and safety people become more involved in painting, I'm sure that robotics will look even better and better. We are just getting started; their future is unlimited. As experience and demand goes, so will robots. They will almost assuredly become the hand sprayer of the 1980's. In my opinion, painting robots are a "can't miss" proposition and definitely are here to stay.[23]

Clearly then, the robotic potential for paint finishing is fully appreciated, especially in the automotive industry where this activity has been very labor-intensive.

Assembly

Unlike the process industries, where entire operations must frequently be automated at one time (either for control purposes or for gaining the economic benefits), assembly in the distance industries must be automated in logical steps.[24] The combined use of automatic assembly machines and manual assembly is both advisable and customary. Those remaining manual operations on the line are usually either too difficult or too costly to automate. The manual laborers remaining on line usually attend to the machines or function as visual inspectors. However, as the assembly process is periodically reviewed with improved knowledge of automatic assembly, the remaining manual operations can be analyzed and considered not only for automation but, more importantly, for cybernation.

To paraphrase Prenting and Kilbridge, automation of assembly is the ultimate answer for improving productivity. Namely, labor cost can be reduced by two-thirds with one machine which does the work of 20 previous laborers. Machinery pay-off periods of less than one year have been reported.[25] According to another source:

> Revolutionary developments in manufacturing technology are opening the door to a new kind of factory, a factory where, in comparison to present day factories, the work environment is greatly improved, where labor requirements are vastly reduced, where raw materials are more effectively utilized and where, as a result, goods can be manufactured at substantially lower cost. The developments responsible for these advances involve an intensified application of computers to the control and management of manufacturing machines and systems. Such systems are termed "programmable" because they can be easily changed to accommodate different products or styles or, in fact, can even be used on a completely different application.[26]

That automated assembly is the wave of the future is clearly indicated from the quotations above. Two distinct concepts for its application are emerging. The first utilizes the substantial memory of the robot to build complete assemblies at a work station and duplicate these work stations as required. The second involves the installation of robots of limited intelligence on an assembly project at each work station. The method chosen will depend on the size and weight of the product to be assembled, the volume in any given lot, and the quality of components from which the assembly is made.

There is no question that advanced cybernetic applications are in place for many production-related activities in the auto industry. Given the industry's significant problems of productivity and competition, it is not an exaggeration to conclude that the cost-reducing and quality control characteristics of robotics make their use imperative. Further, cybernation technology can be expected to become more pervasive and sophisticated in the future. This being the case, we can turn to the issue of how cybernation will affect Black auto production workers.

4

The Robotization of the Auto Industry and the Vulnerability of the Black Production Worker

The auto industry historically has served as an important source of employment for Blacks entering the labor market because of the industry's high wages and job security. Whether this will remain true in the face of the rapid application of cybernation to production-line functions by American auto makers, or whether other work will be available, are major public policy questions.

The Ad Hoc Committee argued that cybernation would modify the mode of production, causing massive and long-run unemployment in the United States if governmental policies did not counteract this tendency. The Committee predicted that such negative effects would fall disproportionately upon Blacks. As has been shown in chapter 3, cybernation is being applied to automotive production, particularly at the assembly plant level. Thus, an obvious question is whether Black production workers in the auto industry are vulnerable to short- and long-run loss of employment opportunities as the Ad Hoc Committee anticipated. It is these questions that will be addressed in this chapter.

The Economic Significance of the Auto Industry to the Black American

The American automobile industry has benefited the Black American workers in many respects. It has served as a main point-of-entry into industrial jobs. Because of its strong union (the UAW), employment security has traditionally been great, and the wages in the auto industry have been higher than those in other industries.

For instance, Herbert R. Northrup has written that "The automobile industry provides the classic example of how the Negro has fared in a dynamic, volatile, yet expanding mass production industry. The extent of Negro employment in the industry is closely related to its growth. . . ."[1]

If we use the 1960s as a benchmark decade, it becomes clear that Blacks benefited most from the increased employment that followed the surge in automobile sales. According to Northrup:

From a depressed area with a labor surplus and unemployment in excess of 16 percent, Detroit turned into a highly prosperous one, importing workers to overcome a critical labor shortage. Anyone willing and able to work could find it—and with earnings which in 1966 averaged about $150 per week.[2]

That the auto industry had become a point-of-entry into the industrial sector for the Black American is made even clearer from the following:

The Detroit area, in particular, saw Negroes emerge as the dominant group in production in many plants. A tour of blue collar employment offices in Detroit any time since 1965 would reveal very few white applicants except at the time of school closings, or summer vacation, or unless the applicant was qualified for a skilled trade. In other parts of the country also Negro employment in the industry rose faster than total employment.[3]

There are high concentrations of Blacks in urban areas which are centers for automobile production. For example, three states—Michigan (30%), Missouri (12%), and Ohio (10%)—account for more than one-half of all cars produced by the American auto industry (see table 4-1). The overwhelming proportions of Blacks residing in these states are located in Standard Metropolitan Statistical Areas which contain the auto production facilities. As table 4-2 indicates, the percentages in Michigan, Missouri, and Ohio are 89, 96 and 61 respectively, distributions which have allowed Black Americans with low skills and minimal education to gain access to high industrial wages. Herbert R. Northrup observed in 1966 that, "It is doubtful if Negroes have so large a share of production jobs in any other major industry."[4]

The attractiveness of the wages earned in the auto industry is apparent in table 4-3. A comparison of the Big Three's annual average weekly salary between 1970 and 1980 to those of the entire motor vehicle and equipment industry, total manufacturing workers, and nonsupervisory workers, reflects the advantage. The wages in the Big Three have outpaced the other wage categories cited in the table. Given these very high wages in the Big Three, we conclude that—in terms of employment and economic significance—the auto industry has been of extreme importance to the Black American. Yet owing to the potential impact of cybernation technology on the auto industry's mode of production, the possibility of either eroding or eliminating those benefits exists.

The Black Production Worker and the Question of Occupational Vulnerability and Employment Risk

Since the mid-1970s the continued availability to Blacks of automotive production jobs and high wages has become increasingly problematic. As the auto industry seeks to improve both its competitive position and profit levels through the adoption of cybernated production techniques, the jobs eliminated will disproportionately

Table 4-1.   Selected Motor Vehicle Assembly by State for Year 1978

State	1978 Model Year Car Assemblies	Percentage Of U.S. Total
Michigan	2,697,709	30.2
Dearborn	174,165	1.9
Detroit	500,149	5.6
Flint	375,998	4.2
Hamtramck	291,533	3.3
Lansing	359,825	4.0
Pontiac	302,750	3.4
Wayne	273,084	3.1
Willow Run	259,516	2.9
Wixom	160,689	1.8
Missouri	1,067,356	12.0
Kansas City	211,082	2.4
Leeds	203,957	2.3
St. Louis	652,317	7.3
Ohio	878,754	9.8
Avon Lake	42,008	0.4
Lorrain	264,023	2.9
Lordstown	333,337	3.7
Norwood	239,386	2.8

Source: MVMA Motor Vehicle Facts and Figures, 1979, p. 19.

affect Blacks. This is not to say that large numbers of White production workers will escape negative impacts. However, Blacks will be harder hit for several reasons.

The jobs most likely to be eliminated by industrial robots are those held by operatives and laborers, or unskilled and semiskilled jobs. Three-quarters of all Blacks employed by the Big Three (in comparison to just over one-half of the White work force) are in these two categories. Thus, a Black employee's job is at relatively higher risk. Conversely, the ratio of Blacks to Whites in nonproduction or less vulnerable jobs is the reverse. Whites are both absolutely and relatively better represented in these positions (nonproduction jobs) and presumably will continue to be.

A person's "occupational vulnerability" is related to the likelihood that robot

Table 4-2. Spatial Distribution of Auto Assembly Plants by SMSAs and Locational Significance of Black Workers

SMSA's By State	No. Blacks By State	No. Blacks in SMSA's With Major Auto Assembly Plant	No. Blacks Near Major Auto Assembly Plant
Michigan	990,663	889,994	89%
Detroit			
Flint			
Lansing			
Ann Arbor			
Jackson			
Missouri	479,746	467,794	96%
Kansas City			
St. Louis			
Ohio	970,130	641,761	61%
Cleveland			
Akron			
Youngstown			
Warren			
Cincinnati			

*These SMSA figures were taken from Department of Commerce 1970 Decennial data, Characteristics of Population, Parts 24, 27, and 37.

Table 4-3. The Comparative Salaried Advantage of the Black Automobile Worker

Year	ANNUAL AVERAGE WEEKLY SALARY			THE BIG THREE* The Annual Average Weekly Salary For Black Auto Workers		
	Motor Vehicle and Equipment Industry	Total Manufacturing Workers	Non-Supervisory Service Workers	GM	Ford	Ch
1970	$170.07	$133.33	$ 96.66	$190.76	$185.00	$183.00
1971	194.46	142.44	103.06	217.23	205.93	203.90
1972	220.59	154.71	110.85	241.80	247.95	236.12
1973	237.51	166.46	117.29	264.71	266.83	249.40
1974	238.32	176.80	126.00	273.60	269.02	236.14
1975	259.53	190.79	134.67	304.07	272.64	257.06
1976	304.16	209.32	143.52	367.11	330.84	303.36
1977	345.84	228.90	153.45	407.77	392.03	334.96
1978	367.63	249.27	163.67	422.40	398.93	351.74
1979	372.37	269.34	175.27	460.40	424.35	358.43
1980	391.42	288.62	190.71	501.60	469.45	404.39

Source: Employment and Earnings, United States, 1909-1978, U.S. Dept. of Labor
(BLS) Bulletin 1312-11, July 1979, pp. 52, 352 and 799. Plus Supplement
to Employment and Earnings revised Established Data, Aug. 1981, pp. 17,
126 and 285.

*Some of these salary figures had to be extrapolated from data obtained from the UAW Research Department, Solidarity House, Detroit, MI, July 1980.

technology can replicate his or her production activities at a favorable cost to the employer. The term "employment risk" refers to the probability that a person made obsolete by cybernation will be reabsorbed into a comparable or better job. These issues of vulnerability and risk reflect the concerns advanced by the Ad Hoc Committee in 1964, as they argued that there would be a disproportionate social cost of cybernation technology placed upon Blacks unless national policies were adopted to manage the transition to a cybernated society. The major analysis in this chapter is designed to determine the extent of occupational vulnerability and employment risk facing Blacks in the auto industry. The intent is not to make precise projections of how many Blacks will have jobs eliminated through cybernation. Rather, the data will be used to indicate how many Blacks are in positions in which human input can be eliminated or largely reduced through cybernation. In a sense, a worst-case scenario is developed.

By looking at the job classification in the auto industry for which employment data are available, it is possible to identify two broad categories of high and low vulnerability to cybernation in the immediate future. They are:

Production Workers   (high)

Craftsmen (skilled)
Operators (semiskilled)
Laborers (unskilled)
Service workers

Nonproduction Workers   (low)

Officials and Managers
Professionals
Technicians
Sales Office and Clerical

The vulnerability of a worker is determined by whether he or she is a production or nonproduction worker, and the extent to which cybernation technology can be applied to specific tasks involved. For example, a pick-and-place robot can easily replace an unskilled laborer. A computerized and a sensory robot can replace not only a skilled worker, but will probably replace certain nonproduction workers (see figure 4-1).

Having determined that manual labor power and/or dexterity are susceptible to replacement through cybernation technology, it is necessary to examine the occupational distribution of Blacks and Whites within the Big Three. This can be done by using data gathered by the Equal Employment Opportunity Commission. EEOC-1 reports for the years 1977 through 1980 on the employment for Chrysler, Ford, and General Motors are utilized in table 4-4 to determine the distribution

Figure 4-1. A Parallelism between Auto Production Workers and Industrial Robots

Laborers
(unskilled)

Operatives
(semiskilled)

Craftsmen
(skilled)

A Pick-and-Place Robot is the simplest version, accounting for about one-third of all U.S. installations. The name comes from its usual application in materials handling: picking something from one spot and placing it at another. Freedom of movement is usually limited to two or three directions - in and out, left and right, and up and down. The control system is electromechanical.

Industrial Robots all have armlike projections and grippers that perform factory work customarily done by humans. The term is usually reserved for machines with some form of built-in control system and capable of stand-alone operation.

A Servo Robot is the most common industrial robot because it can include all robots described below. The name stems from one or more servomechanisms that enable the arm and gripper to alter direction in midair, without having to trip a mechanical switch. Five to seven directional movements are common, depending on the number of "joints," or articulations, in the robot's arm.

An Assembly Robot is a computerized robot, probably a sensory model, designed specifically for assembly-line jobs. For light, batch-manufacturing applications, the arm's design may be fairly anthropomorphic.

A Sensory Robot is a computerized robot with one or more artificial senses, usually sight or touch.

A Programmable Robot is a servo robot directed by a programmable controller that memorizes a sequence of arm-and-gripper movements; this routine can then be repeated perpetually. The robot is reprogrammed by leading its gripper through the new task.

A Computerized Robot is a servo model run by a computer. The computer controller does not have to be taught by leading the arm-gripper through a routine; new instructions can be transmitted electronically. The programming for such "smart" robots may include the ability to optimize, or improve, its work-routine instructions.

Source: *Business Week*, June 9, 1980, p. 64.

Table 4-4. Number of Employees by Race for General Motors, Ford, and Chrysler for Years 1977 through 1980

Occupations	1977*		1978		1979		1980	
	White	Black	White	Black	White	Black	White	Black
Officials and Managers	90,032	7,267	93,861	5,996	87,468	6,788	76,551	5,676
Professionals	67,244	4,392	72,561	3,953	75,989	4,194	67,302	3,690
Technicians	22,358	1,796	24,519	1,693	24,731	1,719	20,231	1,514
Sales Office and Clerical	84,844	9,718	69,067	9,843	63,046	9,065	54,901	7,707
Craftsmen (skilled)	150,596	10,677	153,796	10,755	143,196	9,695	136,310	9,343
Operatives (semiskilled)	541,742	146,351	560,695	148,005	463,820	113,490	406,511	98,827
Laborers (unskilled)	34,076	8,634	39,632	9,622	28,878	7,118	28,374	8,364
Service Workers	26,609	7,744	26,344	7,117	25,048	6,557	22,549	5,670
Grand Totals	1,017,501	196,579	1,040,475	196,984	912,176	158,626	812,729	140,791

Source: UAW Research Department, Solidarity House, Detroit, MI, July 1980.

*Asians, Indians, etc. have been collapsed into the Black category, especially for the Ford Motor Company.

of Blacks and Whites in eight job categories in which operatives and laborers have the greatest vulnerability to replacement by robots.

In table 4-5, where data are reproduced for 1980 as a representative year for exposition purposes, it shows that close to three-quarters (70 percent) of the 140,792 Black workers in the Big Three were operatives in jobs which are potentially vulnerable to elimination because of cybernation technology. In comparison, 50 percent of the total White work force of 812,729 was in the operative category.

If table 4-6 is considered together with previous comments on the growth and development of robotic technology, it can be argued that, in the short run, 87 percent of Black workers (operatives and laborers) are vulnerable to replacement through robotic technology. If we consider recent advances made in sensory robots, an additional 8 percent of Black workers who are craftsmen could be included in this figure. This would bring 95 percent of the jobs currently held by Blacks into the vulnerable category.

Turning to table 4-7 and desegregating these figures, it can be seen that there are variations among the Big Three in the percentage of Blacks who are vulnerable or at risk to cybernation technology. In 1980, GM employed 90,428 Blacks, of which 63,277 were operatives and 4,127 were laborers. Ford Motor Company employed 30,387 Blacks, of which 20,563 were operatives and 3,888 laborers. Chrysler employed 18,095 Blacks in 1980, of which 14,987 were operatives and 349 laborers. Considering all of the production categories except service (skilled, semiskilled, and unskilled), the occupational vulnerability becomes even more extensive. For GM the respective percentages are 4.6, 70 and 7.4 percent (i.e., a total of 82 percent of all Black production workers at GM in 1980 were at risk). The percentages for Ford are 12, 63.7, and 6.4 (or 82 percent) and for Chrysler, they are 1.9, 82.8 and 3.2—88 percent of all Black production workers were vulnerable to replacement through cybernation technology.

If, as all projections indicate, the auto industry continues to increase its substitution of robots and related technology for human labor, there can be no question that Blacks are at a disproportionate risk of losing their jobs, as they are clustered in those jobs with the greatest replacement potential. This is true both in absolute terms and in comparison with Whites employed in the auto industry.

As was emphasized earlier, an analysis of this type cannot predict in any precise way how many production functions will have robotic and related technologies substituted for human labor. However, it is possible to identify jobs which have the highest potential for human obsolescence and the extent to which Black workers are affected. Thus, even though the extent of such substitutions cannot be determined here, all evidence points to their increase in production activities. In fact, projections by General Motors of its robotic usage for 1990 reinforces this view. As figure 4-2 indicates, GM expects by 1990 to have 5000 robots installed in assembly functions, 4000 in machine loading, close to 3000 in welding, and over

Table 4-5. The Vulnerability of Big-Three Employees to Cybernation Technology, 1980

Occupation	White Number	Black Number	Percentage of all Whites in Occupational Category	Percentage of all Blacks in Occupational Category
NONPRODUCTION WORKERS				
Official and Managers	76,551	5,676	9.4%	4.1%
Professionals	67,302	3,690	8.3%	2.6%
Technicians	20,231	1,514	2.5%	1.2%
Sales Office and Clerical	54,901	7,707	6.7%	5.6%
PRODUCTION WORKERS				
Craftsmen (skilled)	136,310	9,343	16.8%	6.6%
Operatives (semiskilled)	406,511	98,827	50.0%	69.9%
Laborers (unskilled)	28,374	8,364	3.5%	5.9%
Service Workers	22,549	5,670	2.8%	4.1%
TOTAL	812,729	140,791	100.0%	100.0%

Table 4-6. The Short-Run Employment Risk to Black Production
Workers Due to the Cybernation of Production in the Big Three

Production Workers

Occupation	Number	Percent	
Craftsmen (skilled)	9,343	8.0%	
Operatives (semiskilled)	98,827	80.0%	Areas of immediate robotic applications
Laborers (unskilled)	8,364	7.0%	
Service Workers	5,670	5.0%	
Grand Totals	122,304	100.0%	

1000 in painting. In terms of the pace of application, no more than 1000 robots
were anticipated for any one of these functions in 1983.

The data clearly support the conclusion that the potentially negative effects
on Blacks of cybernation technology in the automobile industry are great enough,
in the short run, to require public attention. The same is true for the long-run
implications. As will be seen, the potential exists for a continuing and disproportionate reduction in work opportunities for Blacks.

The Long-Run Adjustment of the Work Force in the Motor Vehicle and Equipment Industry to Technological Change

Projecting the long-run effects of cybernation on the employment risk (i.e., the
ability of Blacks and other production-level workers to obtain work) is a hazard-
ous undertaking. Yet, the issue is of such importance that, perhaps, the question
needs to be put in different terms: Is there evidence to indicate that enough uncer-
tainty exists about the long-term effects of cybernation on employment that it should
be given high priority as a policy question on the national agenda? The remainder
of this chapter will seek to address this question.

It could be assumed that if substantial displacement of Blacks does occur in
the auto industry, they would be reabsorbed into other occupational categories
in that field or find parallel production jobs elsewhere in the economy. Data exist
concerning both of these possibilities.

Table 4-7. The Employment Risk to Black Production Workers in GM, Ford, and Chrysler, 1980

Occupation	GM		Ford		Chrysler	
	Black	%	Black	%	Black	%
Officials and Managers	3,956	4.4%	1,091	3.4%	629	3.6%
Professionals	2,135	2.4%	1,311	4.1%	244	1.3%
Technicians	1,083	1.2%	280	.9%	151	.8%
Sales Office and Clerical	5,402	6.0%	1,794	5.6%	511	2.8%
Craftsmen (skilled)	6,706	7.4%	2,066	6.4%	571	3.2%
Operatives (semiskilled)	63,277	70.0%	20,563	63.7%	14,987	82.8%
Laborers (unskilled)	4,127	4.6%	3,888	12.0%	349	1.9%
Service Workers	3,742	4.1%	1,275	4.0%	653	3.6%
Grand Total	90,428	100.0%	32,268	100.0%	18,095	100.0%

Figure 4-2. General Motors Estimated Application of Robotic Production Technology, 1983–90

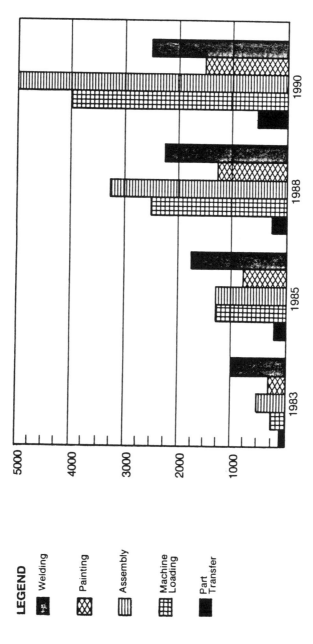

Source: ''Bullish Days in the Robot Business,'' Robotics Age, *The Journal of Intelligent Machines* (September/October 1981), 6–7.

Employment risk was defined earlier as the degree of probability that a person whose job is lost to cybernation would be able to obtain another comparable (or better) position. For people displaced from jobs in the Big Three of the auto industry, an obvious possibility is reemployment by the same firm. In the worst-case scenario developed in the previous section, the statistics are not encouraging that this would happen in the case of Black operatives or laborers.

In 1980 there were 107,191 Blacks in these two categories—76 percent of Black workers at Chrysler, Ford, and General Motors. Even if only a quarter of the group (26,797) was eliminated by cybernation, the number needing to be reabsorbed is almost comparable to the total of all other Blacks (31,719) in nonproduction, service, and craft positions. Thus, unless the number of jobs less vulnerable to cybernation was considerably expanded, or the Black-White ratio in these jobs was significantly altered, few Black production workers would be reemployed in comparable or better positions. The probability is further reduced by the fact that White operatives and laborers would also be displaced (totaling 108,721, if a quarter of the total lost their jobs); and Blacks would have to compete against those workers for reemployment. There can be little doubt that a substantial increase would have to occur in the positions available, as well as the level of skills and training of Black production workers if even a small portion were to be reabsorbed by the Big Three.

The point can easily be made that Chrysler, Ford, and General Motors are not the only options displaced workers will have for employment. Yet, in reviewing a series of Bureau of Labor Statistics (BLS) projections, first through 1985 and then through 1990, there is little cause for optimism. One BLS projection (see figures 4-3 and 4-4) covered employment in the motor vehicle and equipment industry for 1970–85. After a peak in the early 1970s, BLS anticipates a continued decline in total employment, including production workers. BLS comments that production workers will continue to make up about 50 percent of the declining total but notes that

> Many of these workers are engaged in production operations that are relatively labor intensive and have potential for further automation. Semiskilled metal workers (drill press operators, lathe operators, welders, etc.) are expected to decline by 20 percent in response to more widespread use of numerically controlled machines, industrial robots for welding and inspection operations, and more automatic transfer lines.[5]

Looking at these projections for the motor vehicle and equipment industry in general (figure 4-4), the only likely prospects for reabsorption are in the sales and managers, officials, and proprietors categories. Even here, however, there are discounting factors. The projections of 27 and 18 percent growth, respectively, are based on 1985 measured against 1970. As figure 4-3 shows, while the number of jobs anticipated may be greater, there is a general decline in the latter years of the period. To the extent that these nonproduction jobs are available, former

Figure 4-3. Employment in the Motor Vehicle and Equipment Industry, 1960-75, and Projection, 1973–85

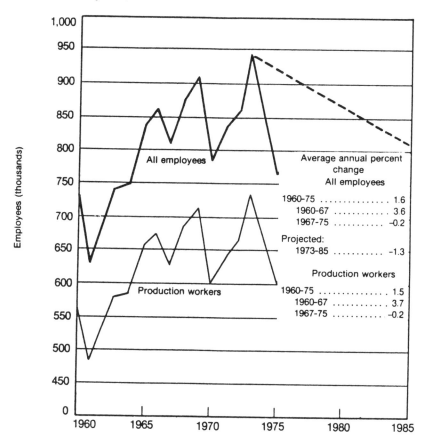

Note: Least squares trend method for historical data; compound interest method for projection.
Source: Bureau of Labor Statistics
Technological Change and Its Labor Implications in Five Industries, Bulletin 1961, 1977, p. 30.

Figure 4-4. Projected Changes in Employment in the Motor Vehicle and
Equipment Industry, by Occupational Group, 1970–85

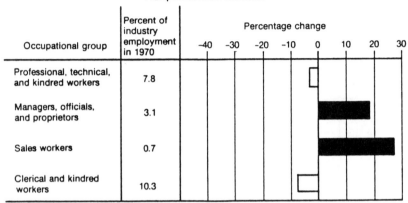

Nonproduction Workers

Occupational group	Percent of industry employment in 1970	Percentage change
Professional, technical, and kindred workers	7.8	
Managers, officials, and proprietors	3.1	
Sales workers	0.7	
Clerical and kindred workers	10.3	

Production Workers

Craft and kindred workers	20.8	
Operatives	49.8	
Service workers	3.0	
Laborers	4.4	

Source: *Technological Change and Its Labor Implications in Five Industries,* U.S. Bureau of Labor Statistics, Bulletin 1961,
1977, p. 32.

production workers seldom move to the managers, officials, and proprietors classification, or even to sales jobs, because of age or lack of education and skills. This is particularly true for Blacks.

The competitive disadvantage of displaced Black production workers is underscored by the BLS. In the categories in which it expects employment to increase by 1985, the BLS forecasts that computer and analytic skills will be in demand. It states that

> Increased use of computers in design, engineering, and production applications should bring about several changes among professional and technical workers and clerical workers. The number of computer specialists (primarily systems analysts and programmers) is expected to increase by 8 percent. Greater use of computer terminals should increase the productivity of drafting technicians and engineers, although the effect of this on employment is unclear. If the volume of work were to remain unchanged, employment might decline. But there is a strong possibility that computer techniques will be used more intensively to improve vehicle design and weight optimization—new analytical work which could absorb people who might otherwise not be needed.[6]

Two other BLS projections reinforce the employment risk argument. In looking at the number of Blacks and Whites in the labor force, the BLS projects that by 1990 the former will exceed the latter (see figure 4-5). Two major points emerge from this data. First, the aggregate labor force is projected to grow. This means additional labor power and dexterity will be available in the face of increasing use of artificially intelligent machines or cybernated production which requires even less human labor. Second, the possibility exists of both massive unemployment and intense racial struggle over the limited number of jobs adaptable to the broad basis of the work force.

The second BLS projection involves the types of jobs that will be available to absorb the growth in the American work force in the future. In the winter of 1981, the BLS made occupational projections for the decade of 1980–90 (see table 4-8). These projections were based on three scenarios: a low, a high (I), and a high (II). Once again, prospects for white-collar employment are good, but the 1980–90 change in blue-collar workers is downwards as a percentage of the work force. Namely, while in 1980, 31.8 percent of the U.S. work force was blue-collar, in 1990 the low trend projection is 31.4 percent, the high trend (I) is 31.7, and the high trend (II) is 31.4 percent. The changes are similar for operatives. In 1980, 13.9 percent of the U.S. work force was blue-collar. In 1990, the low trend will be 13.6 percent, the high trend (I) will be 13.8 percent, and the high trend (II) will be 13.6 percent. These projections are graphically displayed in figure 4-6.

Workers displaced by cybernation technology, especially Black workers, will have to be retrained in order to assume white-collar jobs in the work force. The evidence from the auto industry is inconclusive as to whether the disproportionate number of Black workers to be displaced by cybernation technology will be a short-

Figure 4-5. Projected Growth of Black and White Labor Force and Percentage Change, 1954–90

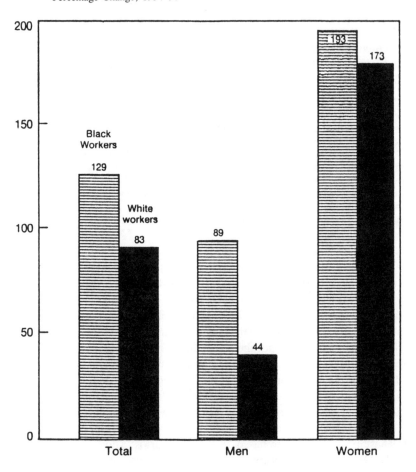

Source: "People and Jobs: A Chartbook of Labor Force, Employment, and Occupational Projections," U.S. Department of Labor, Bureau of Labor Statistics, Regional Report 43, July 1980, p. 9.

Table 4-8. Employment by Major Occupational Group, Actual 1980 and Alternative Projections for 1990

[Numbers in thousands]

Occupational group	1980		1990						Percentage change in employment, 1980-90		
			Low-trend		High-trend I		High-trend II				
	Number	Percent	Number	Percent	Number	Percent	Number	Percent	Low-trend	High-trend I	High-trend II
Total	102,107	100.0	119,591	100.0	127,908	100.0	121,449	100.0	17.1	25.3	18.9
White-collar workers	51,436	50.4	60,755	50.6	64,752	50.7	61,604	50.8	18.1	25.9	19.8
Professional and technical workers	16,395	16.0	19,662	16.4	20,728	16.2	19,917	16.4	19.9	26.4	21.5
Managers and administrators	9,355	9.2	10,563	8.8	11,344	8.9	10,761	8.9	12.9	21.3	15.0
Sales workers	6,822	6.7	8,112	6.8	8,763	6.9	8,205	6.8	18.9	28.5	20.3
Clerical workers	18,864	18.5	22,418	18.8	23,917	18.7	22,721	18.7	18.8	26.8	20.4
Blue-collar workers	32,435	31.8	37,540	31.4	40,497	31.7	36,144	31.4	15.7	24.9	17.6
Craft and kindred workers	12,369	12.1	14,567	12.2	15,756	12.3	14,866	12.2	17.8	27.4	20.2
Operatives	14,206	13.9	16,305	13.6	17,596	13.8	16,487	13.8	14.8	23.9	16.1
Nonfarm laborers	5,860	5.7	6,668	5.6	7,145	5.6	6,791	5.6	13.8	21.9	15.9
Service workers	15,547	15.2	19,103	16.0	20,234	15.8	19,374	16.0	22.9	30.1	24.6
Farm workers	2,669	2.6	2,193	1.8	2,426	1.9	2,327	1.9	-18.5	-9.8	-13.5

NOTE: Due to rounding sums of individual items may not equal totals

Source: "People and Jobs: A Chartbook of Labor Force, Employment, and Occupational Projections," U.S. Department of Labor, Bureau of Labor and Statistics, Regional Report 43, July, 1980, p. 6.

Figure 4-6. Job Growth Scenarios for Major Occupational Categories, 1980–90

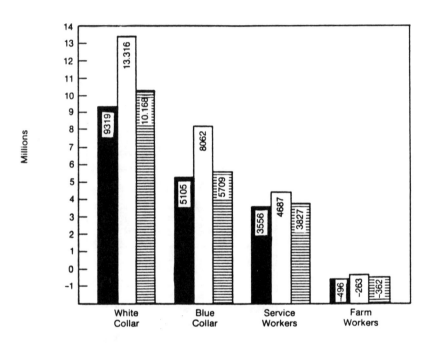

Source: "Three Paths of the Future: Occupational Projections, 1980–90," *Occupational Output Quarterly*, U.S. Department of Labor, Bureau of Labor Statistics, Winter 1981, p. 5.

run phenomenon, and whether those displaced workers be reabsorbed in the long run into the work force. We can now turn to how those groups most central to these issues have responded to the potential job vulnerability and reemployment risk: the automotive industry, auto workers, and the federal government.

5

Public Policy Responses to the Introduction of Cybernation Technology

The data presented in chapter 4 show that, at best, the jobs of a large number of Black automotive workers are being put at risk by cybernation in the short run. At worst, the substitution of robots and related technology for human input in industrial production can lead to creating a permanently unemployable class of people who previously filled semi- and unskilled production jobs. Even so, in a socioeconomic system whereby the means of economic suvival and security are provided through the labor market, the issue of large-scale unemployment due to the obsolescence of human labor has yet to be recognized as a significant public policy question.

For example, while the three major groups of interest in questions of employment and the cybernation of automotive production have different positions, none have placed a priority on making it a matter of public concern. The policy position of the auto industry is relatively clear. Subscribing to a neoclassical economic view, it has assumed that any displaced workers will be reabsorbed into the job market in the long run. Ambiguity is a factor in both the government and UAW perspective. The national government recognizes short- and longer-term displacement, but has pursued policies that assume that—with some public programs—an evolutionary solution will be possible. The UAW is the only one of the three groups that directly represents labor. Yet, the union has not attempted to challenge the introduction of robots on the production line, or to formulate longer-run strategies to ameliorate the effects of cybernation.

Policy Responses of the Principal Actors to the Current Impact of Cybernation Technology in the Auto Industry

In terms of the Ad Hoc Committee-National Commission debate of the mid-1960s, the auto industry can be characterized as viewing the impact of cybernation technology as evolutionary and normal. While the government also views this process as evolutionary and normal, the UAW sees it as revolutionary and heavy. Figure

5-1—which summarizes the positions of these actors in their general policy responses to robotization and how they characterize the effects and the underlying assumptions of their responses—provides a framework for a more detailed discussion.

Management's Responses

> Coming into the 1970's, U.S. manufacturing seemed poised for major advances in automation and—as a result—in productivity. Computers were already well entrenched in the process industries, and they were beginning to improve the efficiencies of such mass production industries as autos and appliances.[1]

This quotation reflects the auto industry's management view that computer-aided manufacturing and computer-aided design will be the means for revitalizing its sagging productivity problem. Management's primary concern and objective is to introduce the labor-saving technology in hopes of increasing productivity, competitiveness, and profits. As far as its responsibility to obsolete workers is concerned, the industry believes this matter should be left to the dynamics of the marketplace or, if necessary, shifted to either the UAW or the federal government. Management's shifting of responsibility, especially for assisting obsolesced or displaced workers, is consistent and logical within the character of market capitalism. The ability of production technology to render portions of the work force obsolete is hardly a new phenomenon. Neither is the assumption that the effects are evolutionary original to this decade.

In 1933, Elizabeth Faulkner Baker conducted a study in which she evaluated changing technology and labor displacement. She found that productivity was increasing and manual labor was being rendered obsolete, but that management showed a genuine lack of concern for those displaced. She felt that its failure to assume direct responsibility for those displaced was based on the theoretical working of the neoclassical model of adjustments in the supply and demand for labor.[2] From a purely rational perspective, bringing cybernation technology into the mode of production to replace human labor power and/or dexterity is technically referred to as "labor saving," or, to quote Walter John Marx: "The very expression 'labor saving' indicates that the machine displaced labor."[3]

Neoclassical economic theory bases management's reaction to the current crisis on the transition from production based on labor power and/or dexterity to cybernation technology. The neoclassical model to which we refer assumes that changes in the mode of production are evolutionary, and that they benefit all of society in the long run. If labor is displaced in the short run, the model postulates, the upgrading of skills for reemployment is necessary. This is the position taken by management. If management can be said to have a policy on this issue of technological change and displacement of workers, it can only be understood within the framework of neoclassical economics to which we now turn.[4]

Figure 5-1. Conceptual Policy Matrix

	Conceptual Policy Matrix		
	Institutional Actors		
	The Auto Industry's Management	Federal Government	The Union UAW
Policy responses to robotization	Supportive with no concern for workers at employment risk	Supportive with some concern for workers at employment risk	Supportive with concern for workers at employment risk
Expected impact of cybernation technology	Evolutionary and normal	Evolutionary and normal	Revolutionary and heavy
Underlying assumptions of the responses of the actors	Neoclassical concept of the economic system Cybernation technology is imperative Short-run unemployment Labor reabsorbed in long run Efficiency criterion is dominant	Keynesian and neoclassical concept of the economic system Cybernation technology is imperative Short-run unemployment Selected assistance to unemployable workers Labor reabsorbed in the long run Efficiency criterion is dominant	Keynesian concept of the economic system Cybernation technology is imperative Labor is not easily reabsorbed in the short run Income security strategy dominates the collective bargaining process, job security strategy is secondary

Figures 5-2 and 5-3 are based on neoclassical theory. In both figures, the wage rates are measured on the vertical axes and the quantity of workers demanded is measured on the horizontal axes. The wage rate coincides with the supply curve in figure 5-2, indicating the firm (the assembly plant) is faced with a perfectly elastic supply curve—the marginal cost of the worker and a union negotiating wage rate. Since the firm's most profitable input is obtained by following the rule that the marginal cost of a factor should equal the marginal revenues product, management will hire Q_{L1} units of labor where D_1 interacts with W_1. Let us assume that management proceeds to introduce the industrial robot. What is the immediate effect on labor? A displacement effect occurs which is measured by $Q_{L2} - Q_{L1}$ and the new demand curve, D_2. What happens to the wage rate, and why is it important?

On the first point, we can turn to figure 5-3, which represents a competitive model and a flexible wage rate. The upward-sloping supply curve reflects the fact that the firm buying labor power will have to offer a higher wage to attract workers from other occupations and localities. However, demand curve D_2 coincides with W_1 and Q_1 and supply curve S_1. Management would argue that the robot increases productivity and the demand for labor. The demand curve D_2 shifts to the right, becoming D_3 which coincides with a higher wage rate and W_3 and Q_3. The firm will hire workers until the supply curve shifts to the right, where S_2 coincides with W_2 and Q_2 (which would represent an acceptable new equilibrium from the perspective of the auto industry and society).

The industry, however, does not uniformly rely on a neoclassical model of a free market to deal with all problems created by cybernation. Management moves outside traditional theory in mapping its responses to highly successful foreign competition which is based on a substantial degree of robotization, as it is in Japan. For example, testimony at various Congressional hearings during the 1970s on the plight of the auto industry reflects a consistent set of themes from management spokespersons. There were a curious but predictable variety of calls for government action—financial aid for Chrysler and protection for the industry against Japanese imports—and for inaction—the elimination of onerous governmental environmental and other regulations which hamper the ability of auto makers to compete.[5]

While it is clearly assumed in these discussions (on the part of both management and the involved Congressional committees) that cybernation would be a key to increased productivity, the impact of the technology of the work force was not an issue raised by the industry. Obviously, the point was that to interfere with the auto manufacturers' ability to substitute capital for labor would reduce their ability to revitalize the industry.

Figure 5-2. Neoclassical Model

Wage Rate

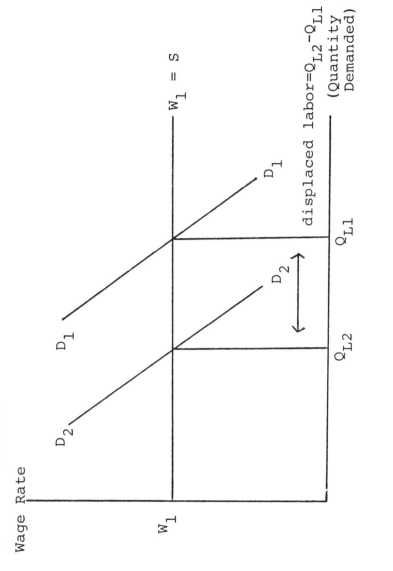

Figure 5-3. Equilibrium Adjustments

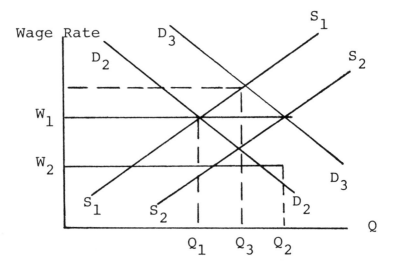

The Federal Government and Employment

The position of the federal government must be considered from two perspectives. One relates to its general policies toward employment. The other concerns its specific responses to the impact of cybernation on employment and the structure of work. As will be shown, there can be a national policy commitment to maintaining adequate levels of employment without the developing of actual programs either directly dealing with the effects of cybernation or even engaging in anticipatory planning. As a result, the government is paradoxically fostering robotization of automotive production but not taking into account the possibility that large numbers of workers may be exiled from the work force because of the long-run work reducing potential of cybernation.

Since the Great Depression, there has been no question that the national interest dictates that the federal government play a role in managing the employment level in the country. However, what role and what policies should be pursued has never been fully resolved. During the 1930s, the government became a massive employer of last resort. From that point to the end of World War II, there was considerable debate over whether full employment should be formally adopted as a national commitment. The Employment Act of 1946 was the outcome of bitter conflict over the issue. In his message of January 21, President Truman asked for enactment of a satisfactory full employment bill; and on February 5 conferees reported a measure entitled the "Employment Act of 1946." It opened with this declaration of policy:

The Congress hereby declares that it is the continuing policy and responsibility of the Federal Government to use all practicable means consistent with its needs and obligations and other essential considerations of national policy, with the assistance and cooperation of industry, agriculture, labor, and state and local governments, to coordinate and utilize all its plans, functions and resources.

1. For the purpose of creating and maintaining, in a manner calculated to foster and promote free competitive enterprise and the general welfare, conditions under which there will be afforded useful employment opportunities, including self-employment for those able, willing and seeking to work, and

2. To promote maximum employment, production and purchasing power.[6]

The bill required the President to submit annually an Economic Report, setting forth current levels of employment, production and purchasing power, together with the levels needed to effectuate the conditions set forth in the policy declaration and a program for carrying out that policy. The bill created a three-member Council of Economic Advisers to assist the President, and a Joint Committee on the Economic Report to consider and report on it.

This Act charges the President with the responsibility for formulating a program to achieve its objectives. The President is assisted by a Council of Economic Advisers. However, while the Act may appear to be a full employment act, it is not. Its aim has been interpreted to be high-level employment; and the content of that concept has changed over time reflecting what appears to be a growing difficulty in reducing the volume of unemployment.[7]

The legislation contained the rhetoric of full employment but not the means to achieve it. In appraising the effectiveness of the Employment Act of 1946 on its 25th anniversary, the Council of Economic Advisers (CEA) noted that the average unemployment rate for the past 25 years had been 4.6 percent but had gone as high as 11.7 percent. Moreover, the Council concluded: "This suggests that we have not appreciably reduced the incidence of small departures from maximum employment but that we have reduced the incidence of large departures, which is just what one would expect aggregate economic policy would do."[8] The CEA believed, according to this report, that for the next 25 years, control of inflation—not unemployment—would be its primary concern.

Despite the difficulty surrounding the issue of effectively addressing unemployment in terms of managing and controlling it, another major bill was introduced in Congress which carried the title "The Full Employment and Balanced Growth Act of 1976." This statute, commonly known as the Humphrey-Hawkins Bill, accepts the concept that the government should be the employer of last resort: if private industry cannot provide employment, then government should.[9]

Once again a national policy to deal with the issue of unemployment became another series of empty phrases. A major obstacle to the successful implementation of the Humphrey-Hawkins Bill, it was argued by opponents, was the cost of training and retraining both the unemployed and unemployable. However, the proponents of this legislation argued that unemployment is also costly. The ineffectiveness of such a policy can be gleaned from the following:

> In 1975, an estimated $75 billion was lost in tax revenues because of high unemployment. Furthermore, about $23 billion was paid out in unemployment compensation benefits for which no productive work was performed. And in addition, a $12 billion tax reduction bill was enacted primarily to stimulate the economy and reduce unemployment. Thus, our present manner of reacting to unemployment involves high costs.[10]

The core of the issue of the Humphrey-Hawkins Bill was whether public policy should be targeted at high employment or low unemployment. These two measures are calculated in different ways. Unemployment data, to a great extent, measure a state of mind, and are compared with the labor force, which tends to be a poorly defined concept. Employment, on the other hand, is much more clearcut and much easier to measure than is unemployment. Measuring unemployment remains a difficult task even today. However, we do not lack a strong national policy insuring adequate levels of employment because its desirability has not been advocated and debated. Rather, the policy of the federal government has been to intervene in the economy regarding employment only in response to market-induced fluctuations (instead of by the establishment of a national plan). Once the corrections are made, it is assumed market forces are the best means for allocating work within the economy.

The most comprehensive policy response in the United States within the last decade to questions of unemployment, underemployment, and retraining is the Comprehensive Employment and Training Act of 1973, or CETA. The Comprehensive Employment and Training Act of 1973 (PL 93–203, as amended) can be summarized thusly:

1. Title 1 establishes a program of financial assistance to state and local governments (prime sponsors) for comprehensive manpower services. Prime sponsors are cities and counties of 100,000 or more, and consortia, defined as any combination of government units in which one member has a population of 100,000 or more. A state may be a prime sponsor for areas not covered by local governments.
2. The prime sponsor must submit a comprehensive plan acceptable to the Secretary of Labor. The plan must set forth the kinds of programs and services to be offered and give assurances that manpower services will be provided to unemployed, underemployed, and disadvantaged persons most in need of help.
3. The sponsor must also set up a planning council representing local interests to serve in an advisory capacity.
4. The mix and design of services is to be determined by the sponsor, who may continue to fund programs of demonstrated effectiveness or set up new ones.
5. Eighty percent of the funds authorized under this title are apportioned in accordance with a formula based on previous levels of funding, unemployment, and low income. The 20 percent not under the formula are to be distributed as follows: 5 percent for special programs to encourage consortia. The remaining amount is available at the Secretary's discretion.

6. State governments must establish a state manpower services council to review the plans of prime sponsors and make recommendations for coordination and for the cooperation of state agencies.

7. Title II provides funds to hire unemployed and underemployed persons in public service jobs in areas of substantial unemployment. Title III provides for direct federal supervision of manpower programs for Indians, migrant and seasonal farm workers, and special groups, such as youth, offenders, older workers, persons of limited English-speaking ability, and other disadvantaged. This title also gives the Secretary the responsibility for research, evaluation, experimental and demonstration projects, labor market information, and job-bank programs. Title IV continues the Job Corps. Title V establishes a National Manpower Commission. Title VI, added in December 1974 under the Emergency Jobs and Unemployment Act. Title VII contains provisions applicable to all programs, such as prohibitions against discrimination and political activity.[11]

CETA clearly fits a neoclassical model. There is little in it that could be called a response to the spectre of large-scale unemployment as a result of (human) labor-obsolescing technology. Rather, it is a short-term policy which will end when the expected new equilibrium is established in the economy.

One of the few federal programs that could be considered to be directed at cybernetically induced unemployment seems to stand the problem on its head. Title II of the Trade Act of 1974 empowers the President to "establish a program of adjustment assistance for communities adversely affected by imports, to expand investment and employment opportunities in such communities, and to improve existing adjustment assistance programs for workers and firms. . . ."[12]

Under this legislation, American auto workers could receive aid if they are adversely affected by competition from Japanese auto makers who are using cybernated production techniques. Ironically, there is no comparable federal legislation which provides similar assistance to a community of auto workers who were put out of work by the massive introduction of production robots in American plants.

The absence of an effective policy on employment at the national level and the lack of a specific program to deal with the potential effects of cybernation reflect a more general unwillingness in America to have the government engage in systematic economic and industrial planning. The steps that are taken by the federal government are ad hoc. The various policies followed are not coordinated and, at times, may be counterproductive. Thus, the government's strong support of revitalizing the auto industry through cybernation does not take the potential displacement of Black workers into account. Organized labor is left as the only possible advocate for a national policy to determine whether the effects will be revolutionary and heavy. Yet, the UAW's position has turned out to be anomalous.

The UAW Response

The UAW, by definition, has the goal of protecting worker income and job security, but finds itself in a perplexing position in relation to cybernation. It has to support the rapid introduction of the cost-effective industrial robot to save the industry. However, this same technology will displace the unskilled and semiskilled production workers that the union is responsible for protecting. The dilemma of the UAW is that if it opposes the short-run displacement of production jobs, the capacity of the auto industry to obtain higher productivity, profitability, and competitiveness in the domestic and international markets is jeopardized. This, in turn, could result in more UAW jobs lost than would be caused by cybernation. As a result, the UAW supports rather than opposes the introduction of robots on the production line.

Given this perspective, absorbing the short-run costs can be rationalized only if it is assumed that, in the long run, a national policy such as that envisioned by the Ad Hoc Committee (see chapter 2) will be adopted. However, the UAW is immobilized on this question: While it doesn't believe that policies based on neoclassical economics are adequate, it has not committed itself and its resources to work for national policies which assume a revolutionary and heavy impact.

Historically the UAW has been more concerned with securing benefits for those working rather than negotiating for guaranteed levels of employment. Even so, the UAW leadership has tended to view labor-saving technology in the work place as revolutionary, and its impact as heavy. Walter P. Reuther—in testimony before a congressional committee in 1955—was eloquent in describing the positive aspects of automation:

> we fully realize that the potential benefits of automation are great, if properly handled. If only a fraction of what technologists promise for the future is true, within a very few years automation can and should make possible a 4-day work week, longer vacation periods, opportunities for earlier retirement, as well as a vast increase in our material standards of living.
>
> At the same time, automation can bring freedom from the monotonous drudgery of many jobs in which the worker today is no more than a servant of the machine. It can free workers from routine, repetitious tasks which the new machines can be taught to do, and can give to the workers who toil at those tasks the opportunity of developing higher skills.[13]

Reuther was equally ready to point out to government that there would need to be adjustments to the new technology by individual workers, industries, and entire communities, and noted that the questions to be answered included: "What should be done to help the worker who will be displaced from his job, or the worker who will find that his highly specialized skill has been taken over by a machine?"[14] Continuing, he warned that industrial leaders are not recognizing the negative consequences of the new technology when "They are the very people who should be in the best position to foresee the difficulties that will have to be met, and in cooperation with government and the trade unions, to take action to meet them."[15]

In 1956, a UAW spokesperson commented that the UAW does not pretend to have all the answers to the problems automation will create for workers. He continued to state that the union is developing a collective bargaining program to

> induce a measure of social responsibility on the part of profit hungry corporations by applying pressure to their pocketbook nerves. Our guaranteed wage agreements are designed primarily to induce social responsibility in the stabilization of employment; but, particularly when further developed in future negotiations, they can become an important factor also in helping to assure a degree of responsibility in the handling of the transition to automation, as well as minimizing hardships resulting from automation.[16]

During the following decade, growth in the economy and the strong bargaining position of the UAW significantly reduced the impact of automation that was forecast in the mid-1950s. Little more was done by the union than by industry or government to anticipate the implications of more sophisticated and powerful forms of automation that would be presented by cybernation.

As it turned out, the ability to protect jobs through the threat of strike is not effective under all circumstances. Several of these circumstances have developed in the 1980s. First, the poor position of the auto industry has led to a poor bargaining position for the UAW. Secondly, the strike as a deciding weapon in the UAW's arsenal will become increasingly insignificant if important aspects of production in the auto industry are able to proceed, effectively and efficiently, without the direct intervention of human beings.

As the industry embarks on its retooling strategy in the 1980s (with the support of government and organized labor), the UAW leadership finds itself in a debilitated collective bargaining position. In response to the serious economic problems of the auto industry itself and the unknowns about the full impact of robotization upon labor, the UAW has been forced to adopt a strategy of concessions to industry (givebacks), early retirement, and job buyouts.

The main objective of UAW strategy during this retooling phase has been to save as many jobs as possible. For with a reduced work force, union dues will shrink and, with them, the union's power. Perhaps that is a possible explanation for management's successful strategy of negotiating concessions from the UAW. David Bensman stated that "Concessions can be extracted only in industries that are being devastated rapidly by such forces as high interest rates . . . or by Government deregulation. . . ."[17]

Bensman's comments might be modified to say that concessions can also be extracted in an industry where manual labor power and/or dexterity is decreasing in importance relative to capital. The following example will illustrate the nature of these UAW policies. Douglas Fraser sought protection for UAW members' jobs at GM in 1981 for eighteen months. In return, the UAW offered to make concessions worth $4800 per worker. The essential points of this agreement were to be:

1. Closing plants and "outsourcing"—buying parts overseas or from other companies—would be restricted during the agreement.

2. All the savings in labor costs would be passed on to consumers. This would be verified by accountants selected by the UAW.

3. All the unionists' sacrifices would be matched by GM's salaried employees, while reductions in supervisory ranks would reflect recent layoffs of production workers.

4. Car dealers would have to reorder one car for every car they sold.

5. If car sales picked up, the UAW could reopen the contract.

6. The union would give back one week paid vacation, nine paid personal days, one bonus Sunday, and the three-percent wage increase due in September, 1982. But before the temporary agreement expired, all benefits and wages would return to the level they would have reached had there been no contract reopening.[18]

GM's labor cost would be reduced by $240.00 per car. The reductions in white-collar pay and dealers' takes increased the figure to $450.00. The UAW argued that GM's sales would increase by four percent and that 15,000 laid-off workers could then be reemployed. However, "GM never agreed to this proposal, and there is no way to tell if it ever would have accepted such modest reductions in labor cost."[19]

The UAW's dilemma is complex. It is fully aware of the need to cybernate the production process in manufacturing autos. In early 1979, a staff paper reported that "Our membership, in general, has favored the introduction of new technological advances, and has recognized them as an essential means of promoting economic progress through increased productivity. Further, labor in general has accepted and encouraged such changes."[20] At the same time, the union's strategies to protect jobs in the short run, let alone the long run, have not matched the current rate of attrition. Wage concessions, early retirements, job buyouts—whatever their immediate utility—do not begin to constitute an adequate policy from the UAW's perspective should the impact of cybernation prove to be revolutionary and heavy. Thus, the UAW doesn't wish to (and probably could not) inhibit the substitution of robots for humans in auto production. Neither is it strongly advocating that the national government assume a responsibility for reassessing the consequences of cybernation for the work force and devising programs to mitigate its negative aspects.

Ironically, then, as the robotization of production becomes more of a reality, there has been no accompanying national debate similar to the one in the 1960s over the possible effects of cybernation on the labor force. In particular, none of the actors most relevant in this study—auto industry management, the federal government, and the UAW—have, in the first two cases, joined the issue. Nor have they adequately pursued the matter in the latter case. As a result, while the basic mode of production in the nation and world is undergoing what many believe is a qualitative shift in the relationship of capital and labor, employment policies remain grounded in what may be obsolete social and economic models.

The issue of the problems of low and semiskilled workers in a cybernated society has found no place on the public agenda. The responses in the auto industry, as described here, are reflective of the nation as a whole. There is wide recognition that production in American industry must become more efficient, that necessary cybernation of production is occurring, and that a more coherent and coordinated governmental policy is required to facilitate this transition. Once higher productivity is achieved, it is assumed that short-term problems of unemployment will be resolved in the market through a new equilibrium.

America's Search for a National Policy

By the late 1970s, a number of initiatives were under way in the United States to formally consider the need for a comprehensive long-term national policy which would allow the economy to recover and insure its steady growth. The Joint Economic Committee (JEC) of Congress, for example, held public hearings in 1978 to take testimony on demographic changes occurring in the nation and to consider their policy implications. The JEC conducted another set of hearings to consider a national economic growth policy in the same year. In 1980, a Congressional hearing was held on a National Employment Priorities Act (H.R. 5040) that was designed to provide federal assistance to individuals and communities affected by plant closings. Also in 1980, a report entitled "A National Agenda for the Eighties" was released by a presidential commission appointed by Jimmy Carter.[21]

One common theme in these activities was the need for coherence and planning at the national level. In the JEC hearings it was observed that ". . . Federal decision making continues to suffer from a lack of integration within both the executive and legislative branches. Problems are tackled in bits and pieces, often producing results favorable in one area but counter-productive in another."[22]

Agreement that systemic changes were occurring in society and in the economy was another theme in these hearings and reports. As it was put by the President's Commission:

Today, as we enter the eighties, the mood is decidedly different. A new constellation of factors—both domestic and international—has arisen in recent years that requires the nation to make some fundamental choices. We no longer have the luxury of recommending more of the same in a variety of areas.[23]

Of the statements that came out of the period, the President's Commission perhaps best represents an awareness of far-reaching changes and a commitment to neoclassic economic assumptions. The Commission does see both economic revitalization and social justice as important issues for the national government, stating:

> —The federal government should provide an overall climate conducive to enhancing the efficient operation of market forces and the private sector. In addition to the measures discussed earlier (e.g., removing barriers needlessly inhibiting market forces), the government should try to foster new relationships of trust and cooperation with commerce and industry.
>
> —The government must persevere in the promotion of social justice and equity for those groups at the lowest end of the socioeconomic spectrum—the poor, the racial and ethnic minorities, the dependent, the handicapped. These considerations are all the more relevant for the coming decade. If public resources are to be heavily committed to the objective of increased economic growth, only a steadfast governmental commitment to social justice will ensure that the available resources, opportunities, and services will be distributed on an equitable basis.[24]

In dealing with these two issues concretely, however, the Commission's logic assumed that evolutionary and normal adjustment of the work force would solve any problems arising from shifts in the mode and location of production. Nowhere is this more obvious than in the Commission's position on "Urban America in the Eighties, Perspectives and Prospects." It is fully accepted that urban America is becoming a postindustrial society. It is also accepted that "in contemporary urban America, both people and places suffer as places transform in step with a changing economy."[25] Yet, the Commission argued that traditional federal policies of aiding declining communities were counterproductive, and that they kept firms and people from choosing optimal locations in terms of productivity for the national economy. As the Commission put it: "Aiding places and local governments directly for the purpose of aiding people indirectly is a policy emphasis that should be re-examined at a time when successes are so few and public resources are so meager?"[26]

Thus, the short-term unemployment produced by cybernation and spatial adjustments by firms would be reduced through long-term spatial adjustments on the part of the work force. It was even proposed that financial aid be provided to workers in order to facilitate their migration from economically declining areas to more profitable ones.[27] The Commission agreed that there were major structural and spatial changes taking place, which while displacing workers in some areas, were necessary to economic revitalization. Yet, the solutions they suggested were a model of a self-correcting economy with facilitative governmental policies. The problem was not seen as one of societal adjustments to far-reaching socioeconomic changes but, rather, as one of simply adjusting the work force to a spatial redistribution of jobs.

What we are faced with, then, at the beginning of a period of significant restructuring of the role of labor in production, is the failure of institutions and political processes to make this issue a matter of public concern and discussion. Neither the most immediately affected parties in the auto industry nor interests in larger society have brought significant attention to the problems involved, particularly for Blacks, in the cybernation of industrial production. The evidence developed in chapters 3 and 4 of the present work clearly supports the view that serious socioeconomic questions exist about the short- and long-term effects of cybernation on the work force. We, as a nation, appear to be assuming these risks without adequately assessing them; and neither the market interplay of capital and labor nor national political processes are providing a public forum for the questions involved.

6

Towards a Public Policy Agenda

Productivity in the American economy has been a matter of wide concern and debate for over a decade. One of the most common views is that, in the industrial sector, physical production methods need to be changed from a labor-intensive mode (utilizing manual labor power and dexterity) to a capital-intensive mode relying on cybernation technology. Nowhere has the loss of productivity been more evident as a problem—and the shift to cybernation technology in the form of robotization seen as the solution—than it has been in the automobile industry. Yet the potential negative effects of cybernation on the work force have received little attention. Moreover, if the impact of this production technology is left entirely to the play of market force, there may be an even greater threat to the socioeconomic fabric of the country than the threat of reduced productivity.

In 1964 the Ad Hoc Committee on the Triple Revolution set out the dilemmas of cybernation for the nation as a whole and, for semi- and unskilled production workers in particular. Members of the committee argued that long-term effects would be revolutionary and heavy unless public policy was utilized to deal with a number of issues ranging from retraining programs for workers whose jobs were lost because of cybernation to how to plan for the transition of our society to one in which income is not dependent upon work.

The problems that the Committee foresaw did not materialize as expected. In retrospect, however, the question can be raised as to whether its analysis or its timetable were correct. Today a variety of indicators show that the extent and rate at which the United States and other industrialized countries are applying cybernation to production more closely fit the circumstances which the Ad Hoc Committee believed required significant public involvement and planning. Paradoxically, the growing evidence that we are moving into an age of cybernation has not been accompanied by a revival of public interest in the issues posed by the Committee.

The present study was undertaken on the assumption that the questions raised by the Ad Hoc Committee are relevant to current conditions and, from a public policy perspective, need to be reassessed, particularly in relation to Black workers in the United States. By looking at the current status of the American auto industry it was possible to examine the potential effects of cybernation on Blacks in the labor force.

Historically the auto industry has been an important point of entry for Blacks into the job market and has provided employment for semi- and unskilled workers with wages that have been among the highest available in industry. However, for the past decade, auto manufacturers have suffered substantial declines in productivity and profits. One solution, widely supported and adopted, has been the application of robotic technology to production functions, with the explicit intent of replacing labor. It is conditions of this type that the Ad Hoc Committee believed would be especially damaging to the ability of Blacks to successfully compete for jobs which are both diminishing in number and becoming more sophisticated in the skills required.

At the beginning of this study three questions were posed. One was whether a qualitative transformation is occurring in industrial production (through the merging of automation and cybernetics) which has the potential to render human labor power and dexterity obsolete in the industrial sector. A review of the evidence showed that this potential exists. The two other questions concerned the future of Blacks in the auto industry in terms of whether there will be displacement in the short run from production jobs, as well as a longer-term inability to obtain equal or better work either in auto-related jobs or the industrial sector in general. As was noted, a number of problems exist in attempting to address these latter questions. One is the lack of adequate time-series data: we are at the initial point in the transition to predominately cybernetic-based production. A second problem concerns the availability of data in general. Neither the auto industry nor the government make available information on the introduction of cybernated production techniques and the immediate effect of such actions on employment. Consequently, rather than directly measuring the effects of cybernation, it was necessary to examine short-run effects in probablistic terms (i.e., the extent to which the jobs of Blacks in the auto industry would be at risk because of the robotization of production, both in absolute terms and in relation to White workers).

After identifying those jobs which are most vulnerable to the substitution of robotic production for human labor power, it was found that the overwhelming majority of Blacks employed in the auto industry are clustered in these positions. While it is true that many Whites are also subject to the same employment risk, a far greater percentage of this group is holding jobs that are not likely to be negatively affected by robotic technology.

The same conclusions were reached upon observing the evidence regarding the longer-term consequences of cybernation technology for Blacks and Whites. Projections, for example, of the Bureau of Labor Statistics to 1990 show that while the percentage of Blacks in the work force will increase, there will be a decrease in the number of semi- and unskilled jobs available. From a more general perspective, it can be argued that when Blacks and Whites are displaced by cybernation technology and are competing for the remaining comparable or better jobs, White workers will have an advantage.

These findings strongly support the conclusions of the Ad Hoc Committee (see chapter 2) that the introduction of cybernation technology into industrial production could potentially create both short- and long-run unemployment for Black workers. At the same time, it must be stated that the current analysis does not settle the question of whether or not these possible effects will occur. What the study does do is validate the argument that, in the absence of the data necessary to resolve the issue of the impact of cybernation, the potential risk is great enough to justify the claim of the Ad Hoc Committee that priority should be given to this question in public discussion and research. Such examination has not yet occurred in the United States.

The parties most directly involved in the large-scale introduction of robots into auto production—the industry's management, the United Auto Workers (UAW), and the federal government—have not sought to make this issue a matter of policy concern. The present failure of public attention to reach even the level it had in the 1960s, when the issue of cybernation was initially raised by the Ad Hoc Committee, is due in part to the immobilization of labor. The UAW has not been able to oppose the substitution of robots for human labor because of the widely accepted view that the American auto industry's competitive position in both the domestic and world market would be destroyed without such action. Further, the decline of the industry has greatly weakened labor's ability to bargain for protection against job displacement.

Auto management, subscribing to a neoclassical economic perspective, assumes that those who suffer short-run unemployment will be reabsorbed in the labor market once a new equilibrium is reached. The national government has supported cybernation in the auto industry and also taken the position that there are no long-term systemic problems. In fact, none of the generally ineffective employment retraining programs recently instituted by the federal government have been initiated in response to a concern over cybernation. Thus, the basic policy problem is not simply to generate the research necessary to acquire an adequate understanding of the potential effects of cybernation on the work force and the structure of society. Rather, there also is a need to gain a place for the issue on the agenda of public concerns.

The importance of this goal can be underscored by combining what is suggested by this study with the projections of the Ad Hoc Committee concerning the results of a failure to implement well-defined public policies for dealing with cybernation. By constructing a worst-case scenario, a number of the potential societal risks of cybernation can be made explicit and provide the basis for discussing policy and research options.

A Worst-Case Scenario

Consider a situation in which large-scale unemployment is created among Black auto workers because of the introduction of robots and other cybernated technology into the industrial sector. Figure 6-1 provides a framework for working through the scenario. Both the workers who lose their jobs in the auto industry and new Black entrants into the work force who cannot find employment, depicted as a_1 and a_2 respectively, aggregate in their community, A. A series of options exist: the unemployed workers can look for employment in the peripheral economy (H); they can seek retraining programs (B); they can try to obtain welfare (C); they might attempt to participate in the underground economy (E); or they could wind up in a penal institution (D). How Blacks are distributed among these possibilities will be significantly influenced by how decision making institutions of the federal government (F) decide to allocate governmental resources.

In a worst-case scenario, the government would invest limited amounts of money in inadequate retraining programs and would be quite generous in providing industry with support for increased cybernation (G). This would mean that more jobs would become vulnerable to cybernation, and more of the available jobs would require skill levels beyond those that most displaced Black workers have or could obtain through retraining programs. Whatever employment could be found would be in the peripheral economy, characterized by small and intermediate size industries which tend to have low wages and unstable employment. This is in contrast to the cybernated core economy (I), with its premium wages, stable employment, and highly skilled work force.

Given this public investment pattern, increasingly large numbers of Blacks would move to welfare, to the underground economy, or to penal institutions, all of which result in dependency and social stigma. In turn, increasing amounts of public money will be required to subsidize welfare, penal institutions and law enforcement. Further, this permanent separation of unskilled and semiskilled workers from wages as their source of income would not be limited to Blacks. The effects of cybernation, in the long run, will fall heavily and disproportionately upon much of what had been the industrial work force, both White and Black. The social and political consequences of this process will lead either to protracted class conflict or to a nation in which significant portions of the population are exiled from the wage economy and relegated to unemployability and second-class status.

A Policy and Research Agenda

The worst-case scenario represents a projection of what is currently happening to the Black auto workers without a national policy to deal with cybernation. Consequently, it becomes an important question whether or not it is possible to address or to reconcile the kind of issues raised by the Ad Hoc Committee. The

Figure 6-1. A Worst-Case Scenario

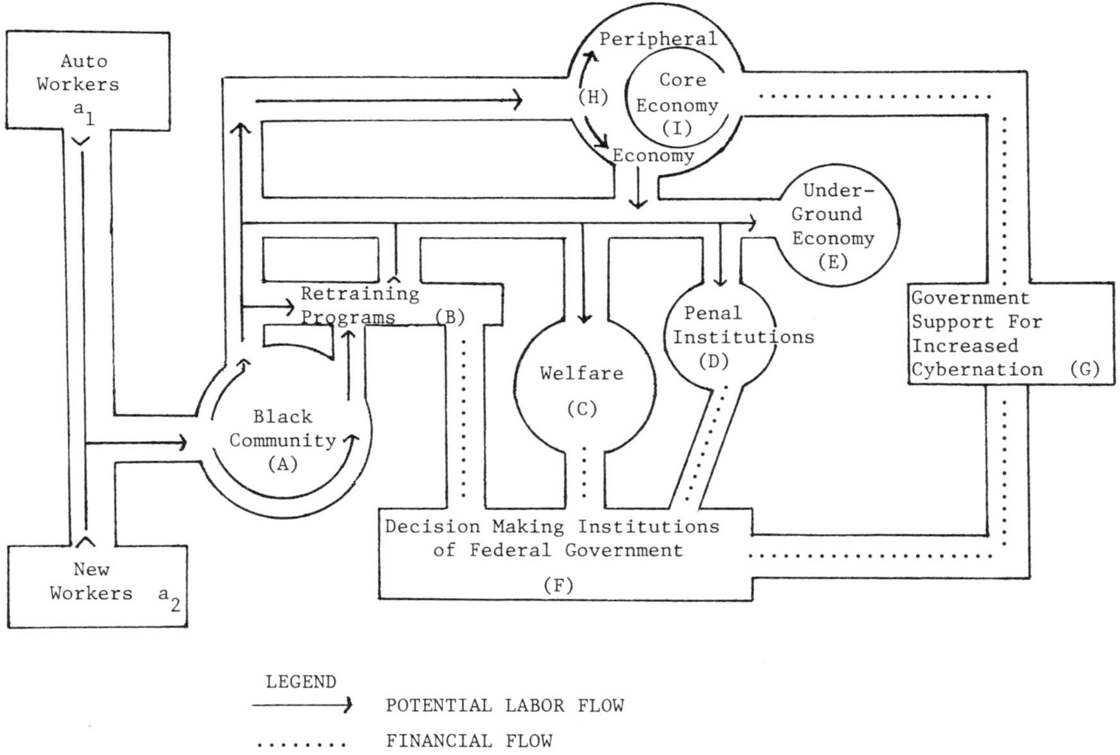

LEGEND

⟶ POTENTIAL LABOR FLOW

········ FINANCIAL FLOW

evidence gathered by briefly reviewing the experience in a selected set of advanced industrial countries with auto industries—Sweden, France, and Japan—suggests that it is possible to deal with this issue.

In Sweden a national consensus has been forged for confronting the issue of advanced technology. The Swedish political system, which has been dominated by a socialist party, has developed the institutional arrangements and priorities to address the issue as an element of national concern.[1]

Regarding the impact of technology on the work force, there is national commitment to lessen the negative consequences. To better understand how this strategy is carried out, mention must be made of the National Labor Market Board (LMB). The LMB was established as a principal means of controlling the economy. This is done, to a large extent, through investment funds, by encouraging investment at the approach of a recession and restricting investment when the economy is buoyant. The LMB motivates the redeployment of workers through training schemes and financial assistance. The task of striking a balance, of facilitating the labor force's adjustment to technological developments, and of trying to forecast changes well in advance are the responsibilities of the National Labor Market Board.[2]

In those cases where labor is displaced because of technological change, the philosophy of the LMB is to retrain and relocate workers while providing adequate financial support. For those workers who will be obsolesced, there exists a policy whose aim is to give the individual government-sponsored courses. Those exposed to seasonal changes are also provided with courses to improve their long-range employment prospects. Finally, as a complement to occupational mobility, a grant is paid to an unemployed worker which is designed to enable him to make an exploratory trip to another town in order to investigate a new job and its environment. Financial assistance is also given to the worker who must visit more than one firm; and the LMB both covers the cost of moving the family and household goods and supplies a resettling allowance.[3]

France has a history, because of its institutional structure, of commitment to national economic and industrial planning. This arrangement is based on a national consensus:

> The French government and the unions have not been content to rely on collective bargaining to protect workers' rights and provide them with benefits. Instead, a vast network of social security provisions offer protection to the general public, including workers. Moreover, certain protections and benefits have been legislated specifically for workers. . . . In France, however, employee protections and benefits frequently have been negotiated by Patronat and the unions first, and then adopted into law with their support. Thus, many of the labor laws constitute the extension of labor agreements to cover the entire economy.[4]

As part of this consensus, economic change, modernization or any process designed to alter the mode of production is subject to worker scrutiny by workers' councils. Further, in places with 50 workers or more, national law requires that workers' councils be provided with

1. Economic rights: general commercial information and annual reports before they go to shareholders; consultation on all general questions, especially those likely to affect efficiency, employment and working conditions.
2. Social rights: consultation on rules and training; advance notice of redundancy and right of appeal to labour inspectorate; advance notice and in certain circumstances agreement on dismissals.
3. Welfare rights: direct or joint administration of institutions and co-decision on employment or dismissal of works' doctors and social workers.
4. Other: representation on statutory safety committees, election of two observers with advisory functions at meetings of management board.[5]

Representatives of the public are also brought into this process of consultation in two ways: (1) through the Economic and Social Council, and (2) through the modernization committees—the vertical and horizontal commissions. The Economic and Social Council is advisory in nature. Its main purpose is to allow the government to obtain the sentiments and reactions of the private economic and social groups to the planning proposals while at the same time providing these groups with the wherewithal to influence the government when constructing the national plan. In other words:

one can view planning in France as a process in which broad economic objectives are determined by the government at the top and broken down by sector and by industry in vertical modernization commissions under the supervision of the planning commission. The detailed industry and sector plans are then aggregated by the horizontal commissions in an attempt to achieve a coherent plan in which all markets are cleared.[6]

Public policy in France is intimately connected with the national plan. The plan tends to reflect the inputs or demands of various social groups. Whether it is automation, cybernation, or robotization, public discussion and assessment of its potential impact begins at the stage of plan construction. It is then that the relationship between the general planning commission and the vertical and horizontal commissions becomes important.

The French worker's best insurance that his case regarding technological impacts will be heard resides within the committees on modernization. These committees are of two kinds: 24 vertical commissions and 5 horizontal commissions. The vertical commissions represent all of the important areas of the economy—manufacturing, steel, chemical, transportation, and the social areas (i.e., housing, public health, culture, and art). Similarly, the horizontal commissions are concerned with the key issues of productivity, manpower, scientific and technical research, economic and financial matters, and regional planning.

To further ensure public review and discussion, each commission has representatives from each of the ministries. In cooperation with the general planning office, the vertical commissions analyze the impact of the proposed national plan on the economy. The French system seeks to offer a coordinating mechanism for including the interests of diverse groups in the consideration of changes in the mode of production; and the impact of these changes is an issue on the public agenda.

Labor-management relations and employment protection policies have long been characteristic of Japanese culture. In Japan, there is reliance on the free enterprise system which is coordinated by government through the Ministry of International Trade and Industry (MITI). There is no question that Japan has developed both a national consensus and an institutional arrangement which recognize and seek to manage the transition to cybernated production through "Reliance on national economic planning which allows the internationalization of externalities or quasi-externalities which enables specific problems created by the growth of one activity to be resolved by a response in another part of the private sector or the public sector."[7]

The success of Japanese economic growth and development is attributable to Japan's unique government-business interaction. The government identifies objectives and priorities for the economy. Also, to facilitate the achievement of these goals, the government attempts to ensure the availability of resources for the private sector. This relationship is not new in Japan. The business community and various government departments have had close relations with each other since the days of the Meiji restoration.[8]

This acceptance by Japanese businessmen, more or less, of the government strategy of setting goals and priorities is based on two very important factors:

1.   The goal of open communication—the desire of government and business to consult each other when considering a major move in the high priority sectors of the economy, and

2.   The goal of consensus formulation—a desire to harmonize differences that may exist within (as well as between) groups.

While Japan's economic planning is not as rigid as that of Sweden or France, it is nonetheless national economic planning. The national economic plans contain, a priori, capital-labor changes resulting from technological advances. The Japanese economic planning agency (EPA) focuses on the national economy, giving industry-by-industry forecasts of proposed production and investment levels.[9] Thus,

> The formulation of the plan involves innumerable meetings of many committees consisting of the E.P.A.'s permanent staff, fulltime delegates from various government departments, business representatives, and a number of university professors. The committees never vote; instead, discussion continues until some sort of consensus is reached. This means that every aspect of the plan is thrashed out in its micro and macro aspects in the context of the domestic and foreign situation, to ensure internal consistency and acceptability to all important participants.[10]

A meaningful discussion of worker protection vis-à-vis labor-management relations has to include a discussion of corporate paternalism. The nature of corporate paternalism is based on the traditional family ideology of mutual affection, benevolence, and reciprocity. The Japanese worker is subject to the following employment practices and/or protections:

1. The institution of lifetime employment—a tenured employee cannot be dismissed from his job unless he is incompetent or has committed a serious offense like theft.

2. The Japanese corporation is organized along a hierarchy of ranks, and young workers are recruited directly out of school. They move up the ladder as they grow older and gain experience.

3. Mandatory retirement—firms want to secure the services of skilled workers for as many years as possible.[11]

In those cases where job obsolescence becomes a factor, workers are shifted to jobs in growing sectors of the economy.

Sweden, France, and Japan differ considerably in their national circumstances but not in their common development of institutional arrangements to plan for and deal with economic and technological change. This is not to say that any of the countries have solved the questions of employment in a cybernated society, or have even adequately addressed them. The important point, however, is that they have recognized the need to establish a means for government, business, and labor to come together in predictable and continuing relationships in order to consider ways of furthering technological innovation and mitigating technology's negative effects on workers and society.

In the United States, a necessary first step is to create a national dialogue concerning the effects of cybernation. The need for introducing new modes of production is already a matter of consensus. However, there is a moral imperative to extend that consensus to the recognition of the necessity of taking precautions against potentially negative effects if the possibility exists that these will fall disproportionately upon any subset of workers. Movement toward this goal will require extraordinary effort on the part of labor and the academic community.

However, creating a new dialogue is only one of the necessary steps. If a national discussion is to be productive there must be a far better understanding both of the effects of cybernation on society and on the work force, and of ways in which undesirable consequences can be reduced or avoided. As this study has shown, we currently do not even collect and organize for policy purposes the limited data that are available relating to cybernation. Thus, there also is a pressing need to establish a research agenda to both foster a national debate on cybernation and inform it.

This agenda should have both a short-run and long-run focus, receiving input from all of the involved elements. The research agenda for the short run should obtain primary and industry-specific data to determine the rate of introduction of cybernetic technology, the rate of manual labor displacement, and the extent of the permanence or irreversibility of the labor displacement process. Secondly, basic data need to be obtained for each sector concerning the availability and kind of jobs. To be successful, the national government will be required to make the generation of such data a matter of national policy and insure their availability to all interested parties.

In the long run, the research should reconsider the issue raised by the Ad Hoc Committee about the nature of the transition to a cybernetic production-based society. How does a system which has been based on the market mechanisms distribute goods and services when the employment-income nexus has been severed? How are people sustained with food, clothing, and shelter when their skills and ability to earn an adequate income have been rendered obsolete? Simulation and Delphi techniques are possible ways to estimate both potential negative impacts under differing assumptions and strategies for their correction. The goal of such research must be to accurately estimate the risks, determine how they are distributed, and assess ways to insure minimum stress and conflict as the mode of production shifts from a reliance on manual labor power and/or dexterity complementing the machine, to a mode of production with a cybernetic base.

For the first time in the history of the capitalist society, the potential exists for the elimination of the employment-income nexus as a requirement for high productivity at extremely low cost. This includes the possibility of rendering obsolete the skills of large subsets of the American people. The Ad Hoc Committee raised these issues 20 years ago. In spite of the fact that there is mounting evidence to support its analysis, the matter has still not been placed on the public agenda as it has in other countries. This study calls for the initiation of a national debate regarding the technical potentialities of cybernetic technology and its social effects. If this research contributes to that objective, the time, resources, and efforts invested in this study will have been worthwhile.

Appendix

Definition of the Terms

The purpose of this section is to clarify the usage of both technical and nontechnical terms in the preceding chapters.

Advanced Cybernetic Machine. This term can refer either to the robot or the analog-digital computers. (Culbertson, J. T., 1963) (Cote, A. J., Jr., 1967) (Bell, D. A., 1962) (Arbib, M. A., 1977) (Simon and Schuster, 1971)

Cybercultural Society. This term tends to describe a future society which would be computer- and information-based. Most decision-making would be "objective" (i.e., computer-based). The traditional notion of work would be supplanted by leisure; paper work would be at a minimum; and most societal activities would be coordinated by "numbers." This kind of society tends to be described in fictional and/or futuristic literature.

Cybernation. This word is a combination of two technical terms, cybernetics and automation. It simply means the computerization of a process. (Parkman, R., 1972) (Michael, N. D., 1962)

The Cybernetic Imperative. This term connotes a strategic need on the part of the industrial sector to introduce the cost-reducing cybernetic machine. It also implies a substantive change occurring in the auto industry, where the production process(es) can be efficiently run without the aid of manual laborers.

Cybernetic Systems. Cybernetic systems study the common properties of different control systems, properties which are quite independent of their material basis. These systems can also be manifested in living nature, in the organic world, and in human collectives. A cybernetic system normally possesses more than one goal. In order to improve its flexibility, it should be able to manipulate these goals so as to order them from most to least appropriate; increase their compatibility; integrate them with information available from sensors and from memory; and use them to initiate commands.

Cybernetics. A very brief (and limited) definition of this term is given by the late Dr. Norbert Wiener: "Cybernetics is the science of communication and control in animals and machines." (Wiener, N., 1948)

Manual Labor Displacement. This term denotes an ongoing process in which the manual labor component of the production process is inevitably replaced by a more "efficient computer." (Parkman, R., 1972) (Sippland Bullen, 1976) (Savas, S. E., 1965) (Jakubi, Kader, Perillo, Automation, 11, 1964) (Neumann, Jon J., 1958)

Manual Labor. This term refers to the human capacity to generate physical activity, particularly in a wage-earning industrial situation. (Marx, K., 1967) (Braverman, H., 1974) (Philip, A., 1890) (Marshall, A., 1893) (Seligrian, R. R. E., 1912)

Robotics. This term deals with the physical action of the robot and the relationship between the robot and the external world. (Young, J. F., 1973) (Culbertson, J. T., 1963)

Notes

Chapter 1

1. Some of the terms used in this study are unfamiliar to many social scientists. The appendix contains a list of terms used in this study.

2. See "Machines Smarter Than Men," in interview with Norbert Weiner, *U.S. News and World Report* (February 24, 1964), 84–86.

3. F. H. George, *Automation, Cybernetics and Society* (New York: Philosophical Library, 1959), p. 1.

4. Ibid., p. 264.

5. John A. Sharp, *Cybernetic Revolution: Its Nature, and Scope* (New Delhi, India: New Book Society of India, 1968), p. 51.

6. Ibid., p. 63.

7. See S. Demczynski, "The Tools of the Cybernetic Revolution," cited in J. Rose (ed.), *Survey of Cybernetics* (New York: Gordon and Breach Science Publishers, Inc., 1969), p. 23.

8. Ibid.

9. Ibid., p. 25.

10. D. N. Michaels, "Cybernation, the Silent Conquest," cited in Richard Romano and Melvin Leiman, *Views on Capitalism* (Beverly Hills, CA: Glencoe Press, 1970).

11. Ibid., pp. 229–33.

12. James Boggs, "The Negro and Cybernation," in *Advancing Technology: Its Impact on Society* (David P. Lauda Company Publishers, 1971), p. 154.

13. U.S. Congress, National Resources Committee Report to the Subcommittee on Social Implications of New Inventions, *Technological Trends and National Policy,* 1937.

14. U.S. Congress, Subcommittee on Economic Stabilization, *Automation and Technological Change: Hearing,* 84th Congress, Sec. 5 (A) of Public Law 304, 79th Congress, 1955, p. 2.

15. See the memorandum by the Ad Hoc Committee entitled, *The Triple Revolution* in the appendix of a book written by Robert MacBride, *The Automated State, Computer Systems as a New Force in Society* (Philadelphia, PA: Chilton Book Company, 1967), pp. 191–207.

16. Ibid.

17. See the 88th Cong., 2nd Sess., January 7–October 3, 1964, Senate Reports Vol. 1–3, U.S. Government Printing Office, Washington, D.C., 1964; and 88th Cong., 2nd Sess., January 7–October 3, 1964, House Reports Vol. 1–3, U.S. Government Printing Office, Washington, D.C., 1964.

18. Gene Bylinsky, "A New Industrial Revolution Is on the Way," *Fortune* (October 5, 1981), 106.

19. "Rebuilding Industry Profitability," *Automotive Industries* (January 1, 1977).

Chapter 2

1. In the course of this study, the author communicated with several of the members of the Ad Hoc Committee. On November 27, 1982, Mr. Ferry (communicating by telephone from Scarsdale, New York, with the author) related the background information on the formation of the Ad Hoc Committee on The Triple Revolution.

2. "Guaranteed Income Asked for All, Employed or Not," *New York Times,* Section 6 (March 22, 1964), 16.

3. U.S. Congress, House, Select Subcommittee on Labor of the Committee on Education and Labor, *The Triple Revolution* Hearing on H.R. 10310, 88th Cong., 2nd Sess., April 14, 15, and 27, 1964, p. 126.

4. Ibid.

5. Ibid., p. 127.

6. Ibid.

7. Ibid., p. 128.

8. Ibid., p. 129.

9. Ibid., pp. 129–30.

10. Ibid., p. 131.

11. Ibid.

12. Ibid., pp. 132–33.

13. Ibid., pp. 133–34.

14. Ibid.

15. For the legislative history on P.L. 88-444 and the formation of the National Commission on Technology, Automation, and Economic Progress see Senate Reports, Vol. 1–3, 88th Cong., 2nd Sess., January 7 and October 3, 1964.

16. House Reports, Vol. 1–3, 88th Cong., 2nd Sess., January 7 and October 3, 1964, U.S. Government Printing Office, Washington, D.C., 1964, pp. 1–2.

17. *Technology and the American Economy,* Report of the National Commission on Technology, Automation, and Economic Progress, Vol. 1, February 1966, U.S. Government Printing Office, Washington, D.C., p. 1.

18. Ibid.

19. Ibid., p. 31.

20. Ibid., p. 67.

21. Ibid., p. 68.

22. Ibid., pp. 109–13.

23. "Delphi Forecasts Predict Changes," Society of Manufacturing Engineers, Dearborn, MI. (June 19, 1978); also see, "The Factory Worker is Economy's Shrinking Man," *Iron Age* (September 22, 1980).

24. "As Robot Age Arrives, Labor Seeks Protection Against Loss of Work," *The Wall Street Journal* (Monday, October 26, 1981), 1.

25. "Bullish Days in the Robot Business," *Robotic Age, The Journal of Intelligent Machines* (September/October 1981), 6–7.

26. Colin Hines, "The Chips Are Down: The Effects of Microprocessors on Employment," *The New Ecologist, Journal of the Post-Industrial Age* 6 (November–December, 1978), 185.

27. Ibid., p. 187.

28. "As Britain Lags Behind in Robotics Race, Some Raise Cry, 'Automate or Liquidate'!" *The Wall Street Journal* (April 5, 1982), 27.

29. "Japan Adopts Robots," *International Management* (June, 1971), p. 55.

30. "New in Japan: The Manless Factory," *The New York Times,* Section 3 (December 13, 1981), 1.

31. "The Robotic Revolution," *Time* (December 8, 1980), 75.

32. "Computers in Manufacturing," *Science,* 215 (February 12, 1982), 818.

33. Colin Norman, "Microelectronics at Work: Productivity and Jobs in the World Economy," in *Worldwatch Paper 39* (October 1980), 8.

34. Ibid., pp. 15–16.

35. Ibid., p. 29.

36. Ibid., p. 54.

Chapter 3

1. U.S. Congress, Senate, Subcommittee on International Finance and Subcommittee on Economic Stabilization of the Committee on Banking, Housing, and Urban Affairs, *The Automobile Industry and World Economy,* Joint Hearing, 96th Cong., 2nd Sess., June 18, 1980, pp. 1–2.

2. Ibid., pp. 4–5.

3. "Japan's Robot Revolution," *World Press Review* (August 1981), 52.

4. "The Growing Use of Industrial Robots—1," *Machinery and Production Engineering* (May 8, 1974), 543.

5. Leo C. Driscoll, "Blue Collar Robots—A Technology Forecast," cited in *Proceedings of the 2nd International Symposium* (Chicago: IIT Research Institute, 1972), p. 197.

6. "Manufacturers Using Robots," *New York Times* (Thursday, October 23, 1980), D2.

7. David A. Van Cleve, "One Big Step for Assembly in the Sky," *Iron Age* (November 28, 1977), 33.

8. Ibid.

9. Paul Kinnucan, "How Smart Robots Are Becoming Smarter," *High Technology* 1, no. 1 (September/October 1981), 32.

10. This discussion on the general applications of robots to production-line activities draws very heavily from William R. Tanner (ed.), *Industrial Robots,* vol. 1 and 2 (Dearborn, MI: Society of Manufacturing Engineers, 1979).

11. Ibid., vol. 2, p. 69.

12. Ibid., vol. 1, p. 24.

13. Ibid., vol. 2, p. 87.

14. Ibid., p. 93.

15. Ibid.

16. Ibid., p. 94.

17. Ibid.

18. Ibid., p. 109.

19. Ibid.

20. Ibid., vol. 2, pp. 149–50.

21. Ibid., pp. 153–56.

22. Ibid., p. 185.

23. Ibid., p. 195.

24. A very rich literature grew up around computerized control of the process industries. See, as one example, Emanual S. Sauas, *Computer Control of Industrial Processes* (New York: McGraw-Hill Book Co., 1965).

25. See for example, T. O. Prenting and M. D. Kilbridge, "Assembly: The Last Frontier of Automation," *Management Review* 55, no. 2 (February, 1965), 7–11. Also, T. O. Prenting, "Automatic Assembly: Economic Considerations," *Mechanical Engineering* 88, no. 8, (August 1966), 30.

26. For the reader unfamiliar with the topic of numerical control, the author recommends the following sources: Charles Ulahos, *Fundamentals of Numerical Control* (Philadelphia: Chilton Book Company, 1968), especially chapters 5, 6, 7, and 10; William E. Miller (ed.), *Digital Computer Applications to Process Control* (New York: Plenum Press, 1965), a publication of Instrument Society of America; "Numerical Control Revolution Sweeps Machine Tool Industry," *Iron Age* 198 (July 7, 1966), 31; E. C. Brown, Jr., "NC Takes on Non-NC Jobs," *American Machinist* 110 (August 15, 1966), 116–17; and L. S. Linderoth, Jr., "Economic Justification for Numerical Control," *Automation* 9 (November 1962), 62–75.

Chapter 4

1. Herbert R. Northrup, *The Negro in the Automobile Industry,* Report No. 1, Philadelphia: University of Pennsylvania, 1968, pp. 34–35.

2. Ibid., p. 35.

3. Ibid.

4. Ibid.

5. *Technological Change and Its Labor Impact in Five Industries,* U.S. Department of Labor, Bureau of Labor Statistics, Bulletin 1961, 1977.

6. Ibid.

Chapter 5

1. "Why the U.S. is Lagging behind in Automation," *Business Week* (June 5, 1978), 62B.

2. Elizabeth Faulkner Baker, *Displacement of Men by Machines* (New York: Columbia University Press, 1981).

3. Walter John Marx, *Mechanization and Culture* (St. Louis: B. Herder Book Company, 1941), p. 10.

4. See for example, Allan M. Cartler, *Theory of Wages and Employment* (Homewood, Ill.: Richard D. Irwin, Inc., 1959), pp. 45–74; Robert L. Crouch, *Macroeconomics* (New York: Harcourt Brace Jovanovich, Inc., 1972), pp. 27–45.

5. U.S. Congress, House, Subcommittee on Economic Stabilization of the Committee on Bank-

ing, Finance, and Urban Affairs, *The Chrysler Corporation Financial Situation, Hearings* on H.R. 5805, Part 1A, 96th Cong., 1st Sess., October 18, 19, 23, and 26, 1979, p. 91.

6. See the Congressional Quarterly 1945–64, Washington, D.C., p. 348.

7. Gordon F. Bloom and Herbert R. Northrup, *Economics of Labor Relations* (Homewood, Ill., 1977), p. 468.

8. Economic Report of the President, transmitted to the Congress, February 1971, p. 21.

9. United States Statutes at Large, 95th Congress of the United States of America 1978 and Proclamations, vol. 92, Part 2, Public Laws 95–473, United States Government Printing Office, Washington, D.C., 1980, Public Law 95–523, October 27, 1978.

10. Bloom and Northrup, *Economics,* p. 469.

11. See, Congressional Quarterly Almanac, 93rd Cong., 1st Sess., 1973, Washington, D.C., pp. 346–54.

12. United States Code Congressional and Administrative News, 93rd Cong., 2nd Sess., 1974, vol. 4, Legislative History, St. Paul, Minn.: West Publishing Co., 1974, p. 7187.

13. "The Impact of Automation," Walter P. Reuther, testimony before the Subcommittee on Economic Stabilization of the Joint Committee of the Economic Report of the United States Congress, October 17, 1955, p. 9.

14. Ibid., p. 10.

15. Ibid.

16. "Practical Approaches to the Problems raised by Automation" by Nat Weinberg, The Archives of Labor History and Urban Affairs University Archives, Wayne State University, 1956, p. 7.

17. David Bensman, "Labor's Painful Dilemma," *Commonwealth* (March 26, 1982), 174.

18. Ibid., p. 175.

19. Ibid.

20. "A View of the UAW Stand on Industrial Robots," a staff paper by Tom Weekley, UAW International Solidarity House, Detroit, January 10, 1979, p. 1.

21. Special Study on Economic Change: A Review of the Panel Meetings May 31 to June 22, 1978, Report of the Joint Economic Committee, Congress of the United States, September 29, 1978, U.S. Government Printing Office, Washington, D.C., 1978; and "A National Agenda for the Eighties," Report of the President's Commission for a National Agenda for the Eighties, U.S. Government Printing Office, Washington, D.C., 1980.

22. JEC report of 1978, p. 114.

23. "A National Agenda for the Eighties," p. 1.

24. Ibid., pp. 13–14.

25. "Urban America in the Eighties Perspectives and Prospects," President's Commission for a National Agenda for the Eighties, p. 4.

26. Ibid., p. 4.

27. Ibid.

Chapter 6

1. As a historical note regarding the Swedish consensus to cooperative national planning see, Walter Korpi, "Industrial Relations and Industrial Conflict: The Case of Sweden," Institute for Research

on Poverty Discussion Papers, University of Wisconsin, Madison, Wisconsin, 1977, pp. 8–21; and Marquis W. Childs, Sweden: *The Middle Way* (New Haven: Yale University Press, 1947), pp. 66–76.

2. Martin Schnitzer, *The Economy of Sweden* (New York: Praeger Publishers, Inc., 1970), pp. 18–20; and Frederic Fleisher, *The New Sweden: A Challenge of a Disciplined Democracy* (New York: David McKay Co., Inc., 1966), chapter 7.

3. H. G. Jones, *Planning and Productivity in Sweden* (Totowa, New Jersey: Rowman and Littlefield, 1976), pp. 39–46; and *Doing Business in Sweden* (Price Waterhouse and Co., 1980), pp. 26–27.

4. Thomas Kennedy, *European Labor Relations* (Lexington, Mass.: D. C. Heath and Company, 1980), p. 52.

5. Innis Macbeath, *The European Approach to Worker-Management Relationships* (London, England: British-North American Research Association, 1973), p. 74. Also, a useful source for guiding this discussion is Stephen S. Cohen, *Modern Capitalist Planning: The French Model* (Berkeley, CA: University of California, 1969).

6. Gary M. Pickersgill and Joyce E. Pickersgill, *Contemporary Economic Systems, A Comparative View* (Englewood Cliffs, New Jersey: Prentice-Hall, Inc., 1974), p. 112.

7. *Japan: Economic and Social Studies in Development,* edited by Neide and Vdo Ernst Simonis (Hamburg, Germany: The Institute of Asian Affairs, 1974), pp. 239–64. Also see Ryutaro Komiya, "Planning in Japan," in Morris Borstein, ed., *Economic Planning: East and West* (Cambridge, Mass.: Ballinger, 1975).

8. Eugene J. Kaplan, *Japan: The Government-Business Relationship, A Guide for the American Businessman,* Bureau of International Commerce, U.S. Department of Commerce, U.S. Government Printing Office, 1972, p. 13.

9. William V. Rapp, "Japan's Industrial Policy," in I. Frank, ed., *The Japanese Economy in International Perspective* (Baltimore: Johns Hopkins University Press, 1975); Ira C. Magaziner and Thomas M. Hout, "Japanese Industrial Policy," Policy Studies Institute No. 585 (London, England, 1980); and William V. Rapp, "Japan: Its Industrial Policies and Corporate Behavior," *Columbia Journal of World Business* (Spring, 1977).

10. *Japan: Economic and Social Studies in Development,* Hamburg, Germany, p. 247.

11. Robert E. Cole, *Japanese Blue Collar* (Berkeley, CA: University of California Press, 1971), pp. 171–224; and Ronald Dore, *British Factory, Japanese Factory* (Berkeley, CA: University of California Press, 1973).

Bibliography

Books

Apter, Michael J. *Cybernetics and Development.* New York: Pergamon Press, 1966.

Automation in Developing Countries. Roundtable Discussion on the Manpower Problems Associated with the Introduction of Automation and Advanced Technology in Developing Countries. Geneva: ILO, 1972.

Baran, Paul, and Sweezy, Paul. *Monopoly Capital.* New York: Monthly Review Press, 1966.

Baranson, Jack. *Technology and Multinationals.* Lexington, Mass.: Lexington Books, 1978.

Barkin, Solomon. *Technical Change and Manpower Planning: Coordination at Enterprise Level.* Paris: OECD, 1967.

Barron, Iann, and Curnow, Ray. *The Future with Microelectronics.* London: Frances, Pinter Ltd., 1979.

Bauer, Raymond A.; Rosenbloom, Richard S.; Sharpe, Laure et al. *Second-Order Consequences: A Methodological Essay on the Impact of Technology.* Cambridge, Mass.: MIT Press, 1969.

Bazezinski, Zbigniew. *Between Two Ages: America's Role in the Technotronic Era.* New York: Viking, 1970.

Bell, Daniel. *The Coming of Post-Industrial Society: A Venture in Social Forecasting.* New York: Basic Books, Inc., 1973.

Best, Fred, ed. *The Future of Work.* Englewood Cliffs, N.J.: Prentice-Hall, 1973.

Bhalla, A. S., ed. *Technology and Employment in Industry.* Geneva: ILO, 1975.

Brady, Robert A. *Organization, Automation and Society: The Scientific Revolution in Industry.* Berkeley: University of California Press, 1963.

Braveman, Harry. *Labor and Monopoly Capital: The Degradation of Work in the Twentieth Century.* New York: Monthly Review Press, 1974.

Bright, James R. *Automation and Management.* Boston: Harvard University Graduate School of Business Administration, 1958.

Centron, Marvin J.; Bartocha, Bodo; and Ralph, Christine A. *The Methodology of Technology Assessment.* New York: Gordon and Breach, 1972.

Chadwick, L. S. *Balanced Employment.* New York: Macmillan Company, 1933.

Chase, Stuart. *Men and Machines.* New York: The Macmillan Company, 1929.

Corey, Lewis. *The Decline of American Capitalism.* New York: Couici Priede Publishers, 1934.

Cowdrick, Edward S. *Industrial History of the United States.* New York: The Ronald Press Co., 1923.

Dahlberg, Arthur. *Jobs, Machines and Capitalism.* New York: The Macmillan Press, 1932.

Dahrendorf, Ralf. *Class and Class Conflict in an Industrial Society.* Stanford, CA: Stanford University Press, 1959.

Diebold, John. *Automation: The Advent of the Automatic Factory.* Princeton, N.J.: D. Van Nostrand Company, Inc., 1952.

Drucker, Peter F. *Managing in Turbulent Times.* New York: Harper and Row, 1980.

Dyke, R. M. *Numerical Control*. Englewood Cliffs, N.J.: Prentice-Hall, Inc., 1967.

Ellul, Jacques. *The Technological Society*. New York: Vintage Books, 1964.

Employment and Technology. Reported by the TUC General Council to the 1979 Congress. London: Trade Union Congress, 1979.

Employment Problems of Automation and Advanced Technology. Proceedings of conference held at Geneva by the International Institute of Labor Studies, 19–24. July 1964. New York: St. Martin's Press, 1966.

Etzioni, Amitai. *The Active Society*. New York: The Free Press, 1968.

Evans, Christopher. *The Micro Millennium*. New York: Viking Press, 1980.

Ferkiss, Victor C. *Technological Man: The Myth and Reality*. New York: George Braziller, 1969.

Frankel, Charles, ed. *Controversies and Decisions: The Social Sciences and Public Policy*. New York: Russell Sage Foundation, 1976.

Georgakas, Dan, and Surkin, Marvin. *Detroit: I Do Mind Dying*. New York: St. Martin's Press, 1975.

Ginzburg, J. A.; Lekhtman, J. Y. A.; and Malou, U. S. *Fundamentals of Automation and Remote Control*. New York: Pergamon Press, 1964.

Godschalk, David R., ed. *Planning in America: Learning From Turbulence*. Washington, D.C.: American Institute of Planners, 1974.

Gottheil, Fred M. *Marx's Economic Predictions*. Evanston, Ill.: Northwestern University Press, 1966.

Gourvitch, Alexander. *Survey of Economic Theory on Technological Change and Employment*. New York: Augustus M. Kelley, 1966.

Grabbe, Eugene M. *Automation in Business and Industry*. New York: John Wiley and Sons, Inc., 1957.

Graham, Otis. *Toward a Planned Society: From Roosevelt to Nixon*. New York: Oxford University Press, 1976.

Gray, Elizabeth; Gray, David D.; and Martin, William F. *Growth and Its Implications for the Future*. Bramford, Conn.: Dinosour Press, 1975.

Haber, William et al. *The Impact of Technological Change: The American Experience*. Kalamazoo, Michigan: The W. E. Upjohn Institute of Employment Research, 1963.

Harris, Donald J. *Capital Accumulation and Income Distribution*. Stanford, CA: Stanford University Press, 1978.

Hays, Samuel P. *The Response to Industrialism 1885–1914*. Chicago: University of Chicago Press, 1957.

Hill, Christopher T., and Utterback, James M., eds. *Technological Innovation for a Dynamic Economy*. New York: Pergamon Press, 1979.

Hines, Colin, and Searle, Graham. *Automatic Unemployment*. London: Earth Resources Research, 1979.

Hunnius, Gerry et al., eds. *Worker's Control: A Reader on Labor and Social Change*. New York: Vintage Books, 1973.

The Impact of Microelectronics on Employment in Western Europe in the 1980's. Brussels: European Trade Union Institute, 1980.

Jackson, Philip C., Jr., *Introduction to Artificial Intelligence*. New York: Petrocelli Books, 1974.

Jaffe, A. J., and Froomkin, Joseph. *Technology and Jobs: Automation in Perspective*. New York: Praeger, 1968.

Jenkins, Clive, and Sherman, Barrie. *The Collapse of Work*. London: Eyre Methuen, 1979.

Jerome, Harry. *Mechanization in Industry*. New York: Publication of the National Bureau of Economic Research No. 27, 1934.

Kidd, Charles V. *Manpower Policies for the Use of Science and Technology in Development*. New York: Pergamon Press, 1980.

Kirkland, Edward C. *A History of American Economic Life*. New York: Appleton-Century-Crofts, 1969.

Klir, Jiri, and Nalach, Miroslau. *Cybernetic Modeling*. Princeton, N.J.: D. Van Nostrand Company, Inc., 1967.

Labor and Automation: A Discussion of Research Methods. Geneva: International Labor Office, 1964.

Labor and Automation: A Tabulation of Case Studies on Technological Change: Economic and Social Problems Reviewed in 160 Case Studies. Geneva: ILO, 1965.

Labor and Automation, Manpower Adjustment Programmes: I. France, Federal, Republic of Germany, United Kingdom. Geneva: ILO, 1967.

Labor and Automation: Manpower Adjustment Programmes: II. Sweden, USSR, United States. Geneva: ILO, 1967.

Labor and Automation: Manpower Adjustment Programmes: III. Canada, Italy, Japan. Geneva: ILO, 1968.

Leon, P. *Structural Change and Growth in Capitalism.* Baltimore: Johns Hopkins Press, 1967.

Levitan, Sar A.; Johnson, William B.; and Taggart, Robert. *Still a Dream: The Changing Status of Blacks since 1960.* Cambridge, Mass.: Harvard University Press, 1975.

Mandel, Ernest. *Late Capitalism.* Norfolk, London: Lowe and Brydone, The Ford, 1972.

Mansfield, Edwin. *The Economics of Technological Change.* New York: W. W. Norton and Co., Inc., 1968.

Marcson, Simon, ed. *Automation, Alienation and Anomie.* New York: Harper and Row, 1970.

Marx, Karl. *Capital: A Critique of Political Economy.* New York: International Publishers, 1967.

Mesthene, Emmanuel. *Technological Change.* New York: New American Library, 1970.

Mintzes, Jospeh, ed. *Science and Technology for Development. Organized Labor's Concern.* Washington, D.C.: American Association for the Advancement of Science Publication No. 79-R-7, 1979.

Morris, Phillipson, ed. *Automation Implications for the Future.* New York: Vintage Books, 1962.

Mueller, Eva et al. *Technological Advance in an Expanding Economy: Its Impact on a Cross-Section of the Labor Force.* Ann Arbor: University of Michigan, Institute of Social Research, 1969.

Muller, Ronald. *Revitalizing America.* New York: Simon and Schuster, 1980.

National Economic Planning: Right or Wrong for the U.S.? Washington, D.C.: American Enterprise Institute, 1976.

Newman, Dorothy. *Protest, Politics and Prosperity: Black Americans and White Institutions 1940–75.* New York: Pantheon Press, 1978.

New York State School of Industrial and Labor Relations. *Technological Change and Human Development.* Ithaca, New York: Cornell University Press, 1970.

Nilsson, Mils J. *Principles of Artificial Intelligence.* Palo Alto, CA: Tioga Publishing Co., 1980.

Noble, David F. *America by Design: Science, Technology and the Use of Corporate Capitalism.* New York: Knopf, 1977.

Nora, Simon, and Minc, Alan. *The Computerization of Society.* Cambridge, Mass.: MIT Press, 1980.

Norman, Colin. *Microelectronics at Work: Productivity and Jobs in the World Economy.* Worldwatch Paper 39. Washington, D.C.: Worldwatch Institute, October 1980.

Northrup, Herbert R., ed. *Negro Employment in Basic Industry: A Study of Racial Policies in Six Industries.* Philadelphia, PA: Industrial Research Unit, Wharton School of Finance and Commerce, University of Pennsylvania.

Pekelis, V. *Cybernetics A to Z.* Translated from Russian by M. Samokhvalov. Moscow: MIR Publishers, 1974.

Pollard, Spencer D. *How Capitalism Can Succeed.* Harrisburg, PA: The Stackpole Company, 1966.

Pollock, Frederick. *Automation: A Study of Its Economic and Social Consequences.* New York: Praeger Publishers, 1957.

Pyke, Magnus. *Automation: Its Purpose and Future.* New York: Philosophical Library, Inc., 1957.

Quinn, S. S., and Francis, X. *The Ethical Aftermath of Automation.* Westminster, Maryland: The Newman Press, 1962.

Ramo, Simon. *America's Technology Slip.* New York: John Wiley, 1980.

Recent Economic Change in the United States. Report of the Committee on Recent Economic Changes, of the President's Conference on Unemployment, vol. I and II. New York: McGraw-Hill Book Company, Inc., 1929.

Ringle, Martin, ed. *Philosophical Perspectives in Artificial Intelligence.* Atlantic Highlands, N.J.: Humanities Press Inc., 1979.

Rose, J., ed. *Progress in Cybernetics*. Vol. 1, 2, and 3. New York: Gordon and Breach Science Publishers, 1970.

———. *Survey of Cybernetics*. New York: Gordon and Breach Science Publishers, Inc., 1969.

Rosenberg, Jerry M. *The Computer Prophets*. London: The Macmillan Company, Collier-Macmillan Ltd., 1969.

Rothwell, Roy, and Zegueld, Walter. *Technical Change and Employment*. New York: St. Martin's Press, 1979.

Schmidt, Emerson P., ed. *Man and Society*. New York: Prentice-Hall, Inc., 1938.

Seidenberg, Roderick. *Anatomy of the Future*. Chapel Hill: University of North Carolina Press, 1961.

Sharpe, William F. *The Economics of Computers*. New York: Columbia University Press, 1969.

Shepard, Jon M. *Automation and Alienation: A Study of Office and Factory Workers*. Cambridge, Mass.: The MIT Press, 1971.

Siegfried, Giedion. *Mechanization Takes Command—A Contribution to Autonomous History*. New York: W. W. Norton and Company, Inc., 1969.

Simons, A. M. *Social Forces in American History*. New York: The Macmillan Company, 1911.

Sleigh, Jonathan et al. *The Manpower Implications of Micro-Electronic Technology*. London: HMSO, 1980.

Spero, Sterling D., and Harris, Abram L. *The Black Worker*. New York: Columbia University Press, 1931.

Stephens, James C. *Managing Complexity: Work, Technology and Human Relations*. Washington, D.C.: The University Press of Washington, D.C., 1970.

Stewart, Frances. *Technology and Underdevelopment*. Boulder, Colorado: Westview Press, 1977.

Stieber, Jack, ed. *Employment Problems of Automation and Advanced Technology. An International Perspective*. London: Macmillan and Co., Ltd., 1966.

Task Force on Education and Employment. *Education and Employment: Knowledge for Action*. Washington, D.C.: Acropolis Books Ltd., 1979.

Technical Change and Economic Policy. Paris: OECD, 1980.

Technological Change and Human Development. An International Conference, Jerusalem, April 14–18, 1969. Ithaca, New York: New York State School of Industrial and Labor Relations, 1970.

Technological Change and Manpower in a Centrally Planned Economy. Geneva: ILO, 1966.

Technology, Employment and Basic Needs. ILO overview paper prepared for the United Nations Conference on Science and Technology for Development. Geneva: ILO, 1978.

Twiss, Brian C. *Managing Technological Innovation*. New York: Longman, 1974.

Ulman, Lloyd, ed. *Manpower Programs in the Policy Mix*. Baltimore: Johns Hopkins University Press, 1973.

Usher, Abbott Payson. *A History of Mechanical Inventions*. Cambridge, Mass.: Harvard University Press, rev. ed., 1954.

Vernon, Raymond, ed. *The Technology Factor in International Trade*. New York: National Bureau of Economic Research, 1970.

Weaver, Robert C. *Negro Labor: A National Problem*. New York: Harcourt, Brace and Company, 1946.

Wedderburn, Dorothy, and Crompton, Rosemary. *Workers' Attitudes and Technology*. Cambridge: Harvard University Press, 1972.

Weinstein, Paul A., ed. *Featherbedding and Technological Change*. Boston: D. C. Heath and Co., 1965.

Welford, Alan Traviss. *Ergonomics of Automation*. London: H. M. Stationary, 1960.

Wesley, Charles H. *Negro Labor in the United States, 1850–1925*. 1968.

Wiener, Norbert. *Cybernetics and Science of Control and Communication Processes in Animals and Machines*. Cambridge, Mass.: MIT Press, 1948.

———. *The Human Use of Human Beings*. Boston: Houghton Mifflin Company, 1954.

Williams, William Appleman. *The Roots of the Modern American Empire.* New York: Random House, 1969.

Yates, Raymond F. *Machines Over Men.* New York: Frederick A. Stokes Co., 1939.

Journals

Alt, John. "Beyond Class: The Decline of Industrial Labor and Leisure." *Telos* 28 (Summer 1976): 55–81.

Baer, Weiner. "Technology, Employment and Development: Empirical Findings." *World Development* (February 1976): 121–30.

Baron, Christopher G. "Computers and Employment in Developing Countries." *International Labor Review* (May–June 1976): 329–44.

Beckman, Norman, ed. "Policy Analysis in Government: Alternatives to Muddling Through." Symposium. *Public Administration Review* 37 (1977): 221–63.

Bergmann, Barbara R., and Lyle, Jerolyn R. "The Occupational Standing of Negroes by Areas and Industries." *Journal of Human Resources* 6 (Fall 1971): 411–33.

Bhalla, A. S. "Technology and Employment: Some Conclusions." *International Labor Review* (March–April 1976): 189–204.

Bloch, Herman D. "Some Economic Effects of Discrimination." *American Journal of Economics and Sociology* 25 (January 1966): 11–23.

Bloice, Carl. "The Black Worker's Future under American Capitalism." *The Black Scholar* (May 1972): 14–22.

Bright, J. R. "Does Automation Raise Skill Requirements?" *Harvard Business Review* (July 1958): 85–98.

Burns, Tom R., and Buckley, Walter, eds. "Power and Control: Social Structures and Their Transformation." *Sage Studies in International Sociology* 6, Beverly Hills, CA.

Callahan, Joseph M.; Fosdick, Richard J.; and Bolton, Eric L. "Rebuilding Industry Profitability." *Automotive Industries* (August 1, 1977).

Chiswick, Barry R. "Racial Discrimination in the Labor Market: A Test of Alternative Hypotheses." *Journal of Political Economy* 81 (November–December 1973): 1330–53.

"Controversy over the Humphrey-Hawkins Proposals to Control Unemployment." *Congressional Digest* 55 (1976): 162–92.

"Data Called from Assembly Lines for 'Instant' Quality Control." *Iron Age* (April 2, 1964).

Faunce, William. "Automation in the Automobile Industry: Some Consequences for In-Plant Social Structure." *American Sociological Review* 23 (1958): 401–7.

———. "Automation and the Automobile Worker." *Labor and Industrial Relations Review* 6, 1 (Summer 1958): 71–72.

———. "Automation and the Automobile Worker." *Social Problems* 6 (1958): 68–78.

Gibbons, Michael, and Johnston, Ron. "The Roles of Science in Technological Innovation." *Research Policy* 3 (1974): 220–42.

Gomolak, L. S. "Special Report: Direct Digital Control in Industry." *Electronics* 37 (October 5, 1964): 73–96.

Gordon, Robert A. "Rigor and Relevance in a Changing Institutional Setting." *American Economic Review* 66 (1976): 1–14.

Gotlieb, C. C., and Borodin, A. "Computers and Employment." *Comm. AMC* 15 (1972): 695–702.

Haberler, Gottfried, and Slichter, Summer H. "Five Views on the Murray Full Employment Bill." *Review of Economic Statistics* 27 (3) (August 1945): 106–12.

Haeffner, E. A. "The Innovation Process." *Technology Review* 75 (March/April 1973): 18–25.

Hamsen, Alvin H. "The Technological Interpretation of History." *Quarterly Journal of Economics* 26, 72–83.

Harder, Delmar S. "Automation, A Modern Industrial Development." *Automation* (August 1954): 46–54.

Hardin, Einar. "Computer Automation, Work Environment and Employee Satisfaction: A Case Study." *Labor and Industrial Relations Review* 4 (13) (July 1960): 567.

Harry, Jerome. "The Measurement of Productivity Changes and Displacement of Labor." *American Economic Review Supplement* 22 (March 1932): 32–40.

Hasegawa, Y. "Quality of Working Life and Robots." *Journal of the Society of Instrument and Control Engineers* (Japan) 16 (1) (January 1977): 111–16.

Hawkins, Augustus F. "Full Employment to Meet America's Needs." *Challenge* 18 (5) (1975): 20–28.

―――. "Planning for Personal Choice: The Equal Opportunity and Full Employment Act." *Annals of the American Academy of Political and Social Science* 418 (1975): 13–16.

Hoffman, Kurt, and Rush, Howard. "Microelectronics, Industry, and the Third World." *Futures* 12 (4) (August 1980): 289–302.

Humphrey, Hubert H. "Planning Economic Policy: Interview." *Challenge* 18, 1 (1975): 21–27.

―――. "The New Humphrey-Hawkins Bill: Interview." *Challenge* 19, 2 (1976): 21–29.

―――. "The U.S. Government's Planning Efforts: A Criticism and a Proposal." *World Future Society* Bulletin 9, 5 (1975): 3–10.

Initiative Committee for National Economic Planning for a National Economic Planning System. *Challenge* 18, 1 (1975): 51–54; *Social Policy* 5, 6 (1975): 17–19.

Institute for Alternative Futures (Washington, D.C.). *World Future Society Bulletin* 11, 2 (1977): 15.

Johnson, Harry. "Technological Change and Comparative Advantage: An Advanced Country's View-point." *Journal of World Trade Law* (January–February 1975): 1–14.

Kahn, H. L. "Automation and Employment." *Labor Law Journal* (November 1959): 796–804.

Kasper, Hirschel. "The Asking Price of Labor and the Duration of Unemployment." *Review of Economics and Statistics* (May 1967): 165–72.

Katrak, Homi. "Human Skills, R & D and Scale Economies in the Exports of the United Kingdom and the United States." *Oxford Economic Papers* (New Series) (November 1973): 337–60.

Licklider, J. C. R. "Man-Computer Symbiosis." *Institute of Radio Engineers Transactions on Human Factors in Electronics* 1 (March 1960): 4–11.

Loveridge, Ray. "Business Strategy and Community Culture—Manpower Policy as a Structured Accommodation of Conflict." *Working Paper,* Series No. 146, Birmingham, England: The University of Aston Management Center (July 1979).

Magaziner, Ira C., and Hout, Thomas M. "Japanese Industrial Policy." *Policy Studies Institute* 585 (January 1980).

McCulloch, W. S. "The Brain as a Computing Machine." *Electrical Engineering* 68 (1949): 492–97.

Meek, Ronald L. "Marx's Doctrine of Increasing Misery." *Science and Society* 26, 4 (1962).

Mehmet, O. "Benefit Cost Analysis of Alternative Techniques of Production for Employment Creation." *International Labor Review* (July–August 1971): 37–50.

Melechanik, G. "Social Costs of the Scientific Technical Revolution under Capitalism." *Problems of Economics* (December 1969): 3–22.

Morman, Robert R. "Automation, Dropouts, and Guidance." *Bulletin of the National Association of Secondary School Principals* 38, 295 (November 1964): 87.

Nelson, R. R., and Phelps, E. S. "Investment in Human, Technological Diffusion and Economic Growth." *American Economic Review* (May 1966): 69–75.

Northrop, Herbert R. "Organized Labor and Negro Workers." *Journal of Political Economy* (October 1943): 206–21.

Pack, H., and Todaro, M. "Technological Transfer, Labor Absorption and Economic Development." *Oxford Economic Papers* 21, 3 (November 1969): 395–403.

Pavitt, Keith. "Technical Change: The Prospects for Manufacturing Industry." *Futures* (August 1978): 283–92.

_____, and Walker, William. "Government Policies Towards Industrial Innovation: A Review." *Research Policy,* 5 (1976): 14–97.

Robinson, Joan. "Marx on Unemployment." *Economic Journal* (June–September 1941).

"Robots Rush in Where Mortals Fear?" *Automotive Industries* (December 1, 1977): 122–23.

"Robots Set for Automotive Assault." *Chilton's Automotive Industries* (May 1978).

Rosenberg, Jerry M. "The Impact of Automation on the Labor Movement." *Adult Education* 13, 3 (Spring 1963): 162–67.

Rupp, Erik. "The RKW: A New Approach Towards Technology Transfer: Methods for the Promotion of Innovation in Small and Medium Sized Companies." *Research Policy* 5 (1976): 398–412.

Sen, Amaryu. "Employment, Institutions and Technology: Some Policy Issues." *International Labor Review* (July 1975): 454.

Solo, Robert. "Problems of Modern Technology." *Journal of Economic Issues* (December 1974): 859.

Stobaugh, Robert B. "How Investment Abroad Creates Jobs at Home." *Harvard Business Review* (September–October 1972): 118–26.

Stout, Thomas M. "Economic Justification of Computer Control Systems." *Automatic Control:* 1–9.

Trade and Employment. Special Report No. 30. Washington, D.C.: National Commission for Manpower Policy, 1978.

Ullrian, Joseph C. "How Accurate Are Estimates of State and Local Unemployment." *Industrial and Labor Relations Review* (April 1963): 434–52.

"Unemployment in the United States, 1930–40." *American Economic Association Papers and Proceedings* 30, 5 (February 1941).

Vernon, Raymond. "International Investment and International Trade in the Product Cycle." *Quarterly Journal of Economics* (May 1966): 190–207.

Warner, K. E. "The Need for Some Innovative Concepts of Innovation." *Policy Sciences* 5 (1974): 433-51.

Willhelm, Sidney M. "A Reformulation of Social Action Theory." *American Journal of Economics and Sociology* 26 (January 1967): 23–31.

Williams, T. J. "Computers, Automation and Process Control." *Ind. Eng. Chem.* 56 (1964): 47–56.

_____. "What to Expect from Direct Digital Control." *Chem. Eng.* 71 (March 2, 1964): 97–104.

Winner, Langdon. "The Political Philosophy of Alternative Technology: Historical Roots and Present Prospects." *Technology in Society* 1 (1979): 75–86.

Public Documents

National Academy of Engineering. Committee on Public Engineering Policy: Colloquium. *Perspective on Benefit-Risk Decision Making.* April 26–27, 1971. Washington, D.C., 1972.

_____. *A Study of Technology Assessment.* Report to House Committee on Science and Astronautics. July 1969.

National Academy of Science. *Technology: Process of Assessment and Choice.* Report to House Committee of Science and Astronautics. July 1969.

_____. *Microstructure Science, Engineering and Technology.* Washington, D.C., 1979.

National Commission on Supplies and Shortages. Advisory Committee on National Growth Policy Processes. *Forging America's Future: Strategies for National Growth and Development.* With Appendices, 3 vols. Washington, D.C.: U.S. Government Printing Office, December 1976.

U.S. Congress. U.S. House of Representatives. Committee on Banking, Currency and Housing. *The Automotive Industry and Its Impact upon the Nation's Economy.* Hearings before the Automobile Industry Task Force. 94th Cong., 1st sess., 1975.

_____. *National Growth and Development.* Hearings, 94th Cong., 1st sess. Washington, D.C.: U.S. Government Printing Office, 1975.

U.S. Congress. U.S. House of Representatives. Committee on Education and Labor. *Oversight Hearings on the Impact of the Canadian-American Automotive Agreement on Employment in the U.S.* Hearings before the Subcommittee on Labor Standards. 94th Cong., 1st sess., 1975.

U.S. Congress. U.S. House of Representatives. Committee on Merchant Marine and Fisheries. Subcommittee on Fisheries and Wildlife, Conservation, and the Environment. *Growth and Its Implications for the Future: Effects National Growth Will Have on Resources, the Environment, and Food Supply in the Future.* 3 vols. 93rd Cong., 1st sess. Washington, D.C.: U.S. Government Printing Office, 1973.

_____. *Computers Simulation Methods to Aid National Growth Policy.* 94th Cong., 1st sess. Washington, D.C.: Government Printing Office, 1975.

U.S. Congress. U.S. House of Representatives. Committee on Science and Astronautics. *A Forecast of Technological Trends, Their Societal Consequences and Science Policy Strategies of the Future.* Proceedings of the hearings, by T. A. Gordon, 1970: 429–70.

_____. *Science Policy,* Prepared by the Science Policy Research Division, Congressional Research Service, Library of Congress, April 1972.

_____. *Technical Information for Congress.* Report of the Legislative Reference Service, Library of Congress, July 1969.

_____. *Technology Assessment.* Hearings before the Subcommittee on Science, Research, and Development, November and December, 1969.

_____. *Technology Assessment Seminar.* Proceedings before the Subcommittee on Science, Research, and Development, September, 1967.

U.S. Congress. U.S. House of Representatives. Committee on Science and Technology. *Long-range Planning.* Joint Hearings. 94th Cong., 2nd sess. Washington, D.C.: U.S. Government Printing Office, 1976.

_____. Subcommittee on the Environment and the Atmosphere. *Long-range Planning.* 94th Cong., 2nd sess. Washington, D.C.: U.S. Government Printing Office, 1976.

U.S. Congress. Congressional Budget Office. *Income Disparities between Black and White Americans: A Background Paper.* Washington, D.C.: U.S. Government Printing Office, 1977.

U.S. Congress. Congressional Research Service. *Technology and Trade: Some Indicators of the State of U.S. Industrial Innovation.* Report prepared for Subcommittee on Trade, Committee on Ways and Means, U.S. House of Representatives, April 21, 1980.

U.S. Congress. Joint Economic Committee. *A Full Employment Policy.* By Leon H. Keyserling. 94th Cong., 1st sess., December 10, 1975.

_____. *1976 Joint Economic Report.* Senate Report 94–690. 94th Cong., 2nd sess. Washington, D.C.: U.S. Government Printing Office, 1976.

_____. *Long-range Economic Growth.* Hearings. 94th Cong., 1st sess. Washington, D.C.: U.S. Government Printing Office, 1976.

_____. *National Economic Planning, Balanced Growth, and Full Employment.* Hearings. 94th Cong., 1st sess. Washington, D.C.: U.S. Government Printing Office, 1976.

_____. *Pollsters Report on American Consumers and Businessmen.* Hearings: 94th Cong., 1st sess. Washington, D.C.: U.S. Government Printing Office, 1975–76.

_____. *Productivity.* Hearings before the Joint Economic Committee. 96th Cong., 1st sess., June 5, 6, 1979.

_____. *Soviet Machine Tools: Lagging Technology and Rising Imports, in Soviet Economy in a Time of Change.* By Grant James. October 1979.

_____. *Toward a National Growth Policy: Federal and State Developments in 1974: A Report.* Prepared by the Library of Congress, Congressional Research Service. 94th Cong., 1st sess. Washington, D.C.: U.S. Government Printing Office, 1975.

_____. *Unemployment: Terminology, Measurement and Analysis.* Subcommittee on Economic Statistics. Washington, D.C.: U.S. Government Printing Office, 1961.

_____. *U.S. Economic Growth from 1976 to 1986: Prospects, Problems, and Patterns.* 12 vols. 94th Cong., 1st and 2nd sess. Washington, D.C.: U.S. Government Printing Office, 1976.

U.S. Congress. Office of Technology Assessment. *An Assessment of Information Systems Capabilities Required to Support U.S. Materials Policy Decisions.* Washington, D.C.: U.S. Government Printing Office, December 1976.

_____. *Technology and Steel Industry Competitiveness.* Washington, D.C.: U.S. Government Printing Office, 1980.

U.S. Congress. Senate. Committee on Banking, Housing and Urban Affairs. *Chrysler Corporation Loan Guarantee Act of 1979 S. 1937 to Authorize Emergency Loan Guarantees to the Chrysler Corporation Part I.* Washington, D.C.: U.S. Government Printing Office, 1979.

_____. *Government Regulations of the Automobile Industry.* Supplemental U.S. Government Publications Reports and Research Papers. 96th Cong., 1st sess., 1979.

_____. Subcommittee on Economic Stabilization. *Impact of Government Regulation on the Automobile Industry and Effects on the Overall Economy.* 96th Cong., 1st sess. Washington, D.C.: U.S. Government Printing Office, 1979.

U.S. Congress. Senate. Committee on Commerce. Committee on Government Operations. *Domestic Supply Information Act: Joint Hearings.* 93rd Cong., 2nd sess. Washington, D.C.: U.S. Government Printing Office, 1974.

U.S. Congress. Senate. Committee on Foreign Relations. Subcommittee on Multinational Corporations. *Multinational Corporations and U.S. Foreign Policy.* 93rd and 94th Cong. Washington, D.C.: U.S. Government Printing Office, 1973–76.

U.S. Congress. Senate. Committee on the Judiciary. *Public Participation in Federal Agency Proceedings, S. 2715: Hearings.* 94th Cong., 2nd sess. Washington, D.C.: U.S. Government Printing Office, 1976.

U.S. Congress. Senate. Committee on Labor and Public Welfare. Selected readings in employment and manpower, vol. 6. *History of Employment and Manpower Policy in the United States: Hearings before Subcommittee on Employment and Manpower.* Washington, D.C.: U.S. Government Printing Office, 1965.

U.S. Department of Commerce. *The Social and Economic Status of the Black Population in the United States: An Historical View, 1790–1978.* Current Population Reports. Washington, D.C.: U.S. Government Printing Office, 1979.

U.S. General Accounting Office. *U.S. Actions Needed to Cope with Commodity Shortages.* B-114824. Washington, D.C.: U.S. General Accounting Office, April 29, 1974.

_____. *Long-range Analysis Activities in Seven Federal Agencies.* PAD-77-18. Washington, D.C.: U.S. General Accounting Office, December 3, 1976.

U.S. President. Domestic Council. Committee on Community Development. *National Growth and Development: Second Biennial Report.* Washington, D.C.: U.S. Government Printing Office, December 1974.

_____. *Report on National Growth and Development: The Changing Issues for National Growth.* Washington, D.C.: U.S. Government Printing Office, February 1976.

U.S. President. *Economic Report of the President Together with the Annual Report of the Council of Economic Advisers.* Washington, D.C.: U.S. Government Printing Office, 1973, 1974, 1975, 1976, 1977.

_____. *International Economic Report of the President and Report of the Council on International Economic Policy.* Washington, D.C.: U.S. Government Printing Office, 1973, 1974, 1975, 1976, 1977.

Magazines

"Automatic Control by Distributed Intelligence." *Scientific American* (June 1979).

"Better Parts Handling Spells Bigger Profits (Plastics)." *Plastic World* 33 (October 1976): 30–33.

Boden, M. A. "Social Implications of Intelligent Machines." *Radio and Electronic Engineer* 47, 8–9 (August–September 1977): 393–99.

Christner, Westerly A. "Anything You Can Do a Robot Can Do Better." *Sweden Now* 12, 3 (1978): 14–17.

Conference on Industrial Robots, Lausanne, Switzerland. October 1974. Birkhauser Verlag: Basel Switzerland.

Conway, A. "The Rise of the Robot." *Management Today* (June 1973): 37–38.

Coombs, Rod, and Green, Ken. "Slow March of the Microchip." *New Scientist* (August 7, 1980).

Dangelmayer, David. "The Job Killers of Germany." *New Scientist* (June 8, 1978).

"Eliminating the Human Factor." *MacLean's*, 90, 9 (December 26, 1977).

Engelberger, J. F. "Robots Make Economic and Social Sense." *Atlantic Economic Review* 27, 4 (July/August 1977): 4–7.

Frohlich, G. "Use of Industrial Robots in Sweden." *Foerdernand Heben* (Germany) (Fohban), 26 (13) (October 1976): 17–19.

George, Edwin B. "Should Full Employment Be Guaranteed?" *Dun's Review* 5 (October–December 1947).

Ginzberg, Eli. "The Pluralistic Economy of the U.S." *Scientific America* 235, 6 (December 1976): 25–29.

"GM's PUMA Robot Tackles Assembly Line Chores." *Iron Age* 221 (June 5, 1978): 12–13.

"GMR's Computer Vision Systems: How They Look Today." *General Motors Research Laboratories* 13, 1 (January–February 1978).

Grant, Donald. "Robots Are Automatic Success." *Globe and Mail* (July 12, 1979): 1.

Gregory, Gene. "The March of the Japanese Micro." *New Scientist* (October 11, 1979).

Grosswirth, Marvin. "The Robots Are Coming." *Datamation*, 24, 12 (November 15, 1978): 146–50.

"The Growing Use of Industrial Robots." *Machinery and Production Engineering*, 124, 3206 (May 8, 1974): 542–46.

"High Technology, Wave of the Future or a Market Flash in the Pan?" *Business Week* (November 10, 1980).

Howland, Robert. "Brave New Workplace." *Working Papers* (November–December 1980).

"Japan Stuns the American." *The Economist* (February 23, 1980).

"The Longbridge Robots Will March over the Transport Union." *New Economist* (April 19, 1980).

Marsh, Peter. "Britain Grapples with Robots." *New Scientist* (April 24, 1980).

_____. "Robots See the Light." *New Scientist* (June 12, 1980).

_____. "Toward the Unmanned Factory." *New Scientist* July 31, 1980): 373–75.

"Microelectronic Survey." *The Economist* (March 1, 1980).

Minsky, Marvin. "Telepresence." *OMNI* (June 1980): 45–52.

Nolting, Louran, and Feshback, Murray. "R and D Employment in the U.S.S.R." *Science* (February 1, 1980).

"The Office of the Future." *Business* (June 30, 1975): 48.

"People vs. Machines." *U.S. News and World Report* (January 29, 1973): 44.

Robinson, Arthur L. "Perilous Times for U.S. Microcircuit Makers." *Science* (May 9, 1980).

"Robots Join the Labor Force." *Business Week* (June 9, 1980).

"Robots Reduce Exposure to Some Industrial Hazards." *National Safety News* 110, 82 (July 1974).

Shaiken, Harley. "The Brave New World of Work in Auto." *In These Times* (September 19–25, 1979).

_____. "Detroit Downsizes U.S. Jobs." *The Nation* (October 11, 1980).

Strom, Robert. "Education for a Leisure Society." *The Futurist* (April 1975): 93.

Ward's Automotive Yearbook, 41st edition. Detroit: Ward's Communications, 1972.

"When the Robots Move In." *Business Week* (June 9, 1980).

Wilkinson, Max. "The Coming of the Robot Workplace." *Financial Times* (June 14, 1978).

Reports

Bakke, E. Wight. *The Mission of Manpower Policy*. Kalamazoo, Michigan: Upjohn Institute, April 1969.

Baldwin, George B., and Shultz, George P. *Automation: A New Dimension to Old Problems.* Proceedings of Seventh Annual Meetings of the Industrial Relations Research Association, 1954.

Barney, Gerald O., ed. *The Unfinished Agenda: The Citizen's Policy Guide to Environmental Issues.* Task Force Report. Sponsored by the Rockefeller Brothers Fund. New York: Thomas Y. Crowell, 1977.

Belitsky, Harvey A. *Productivity and Job Security: Retraining to Adapt to Technological Change.* Washington, D.C.: National Center for Productivity and Quality of Working Life, Winter 1977.

Hasegawa, Y. *Industrial Robot Applications for Improving Quality of Working Life.* In the proceedings of the Seventh International Symposium on Industrial Robots, October 19, 1977.

Keyserling, Leon H. *Liberal and Conservative National Economic Policies and Their Consequences, 1919–1979.* Conference on Economic Progress. Washington, D.C., September 1979.

Levine, Louis. *Effects of Technological Change on the Nature of Jobs.* O.E.C.D. Conference. Washington, D.C., 1964.

Newspapers

Burgen, Carl. "The Bulls Are Back in Venture Capital." *Wall Street Journal.* November 17, 1980.

Greenberger, Robert S. "New Civil Servants Sound Off a Lot But Stay in Line: Mail Robots in Federal Offices Do Job Faster, But Union Thinks They Are Inhuman." *Wall Street Journal,* 193:1. April 16, 1979.

Harrington, Michael. "The Productivity Ploy." *New York Times.* March 21, 1980.

Lohr, Steve. "The New Industrial Robots Are Punching Clocks Faster." *New York Times.* News of the Week. 8E, Sunday, April 27, 1980.

Rowland, Mary. "Chrysler Union Says Security Critical Issue." *Evening Journal.* Wilmington, Delaware. September 10, 1979, p. 6.

Udall, Stewart. "The Failed American Dream." *Washington Post.* June 12, 1977.

Other Sources

"Chrysler First U.S. Automaker to Use Unimation Welding System." Source Unknown. 4 pages.

Cole, Ralph I. "Countering Technological Obsolescence: Marshalling Pertinent Forces." Washington, D.C.: The American University. January 1978. (Unpublished paper.)

———. "Development of a High Capability for Technology Transfer in Saudi Arabia." Washington, D.C.: National Science Foundation. January 1979.

"Delegates to UAW White Collar Conference Explore Methods to Deal With New Technology." White Collar Reports. March 16, 1978, pp. A10–16.

Ginsburg, Helen. *Unemployment, Subemployment, and Public Policy.* Center for Studies in Income Maintenance Policy, New York University School of Social Work, 1975.

National Urban League, Inc. *The State of Black America 1977.*

Rosen, Sumner. "Social Policy and Manpower Development." Cited in *Manpower and Employment, A Source Book For Social Workers.* Edited by Margaret Purvime. New York: Council on Social Work Education, 1972, pp. 7–25, 40–65, 189–202.

Smith, James P., and Finis, Welch. "Race Differences in Earnings: A Survey and New Evidence." Santa Monica, CA: The Rand Corporation, 1978.

"U.S. Capitalism in Crisis." A Union of Radical Political Economics Publication, 1978.

Zager, Robert. "Productivity and Job Security: Attrition Benefits and Problems." Washington, D.C.: National Center for Productivity and the Duality of Working Life, Fall 1977.

Index